H.-M. Hoogewoud   G. Rager   H.-B. Burch

Computed Tomography, Anatomy, and Morphometry of the Lower Extremity

With Contributions by
P. Cerutti and G. Rilling

With a Comparative CT and Anatomical Atlas
Containing 48 Plates Consisting of 123 Separate Illustrations,
and 12 Further Figures

Including Morphometry and 3D Graphic Software
on a 5.25″ Floppy Disk for IBM PCs and Compatibles

Springer-Verlag Berlin Heidelberg New York
London Paris Tokyo Hong Kong

ISBN-13:978-3-642-74651-2   e-ISBN-13:978-3-642-74649-9
DOI: 10.1007/978-3-642-74649-9

Library of Congress Cataloging-in-Publication Data. Hoogewoud, H.-M. (Henri-Marcel), 1953- Computed tomography, anatomy, and morphometry of the lower extremity : with a comparative CT and anatomical atlas containing 48 plates consisting of 123 separate illustrations, and 12 further figures / H.-M. Hoogewoud, G. Rager, H.-B. Burch ; with contributions by P. Cerutti and G. Rilling. p. cm. Includes biblio-
graphical references. ISBN-13:978-3-642-74651-2 (U. S.: alk. paper) 1. Extremities, Lower-Anatomy-Atlases.
2. Extremities, Lower-Tomography-Atlases. 3. Extremities, Lower-Measurement. I. Rager, Günter H.,
1938- . II. Burch, Hansbeat. III. Title. [DNLM: 1. Leg-anatomy & histology-atlases. 2. Tomography, X-
Ray Computed-atlases. WE 17 H779c] QM549.H66 1989 611'.98'0222-dc20 DNLM/DLC for
Library of Congress 89-21819 CIP

© Springer-Verlag Berlin Heidelberg 1990
Softcover reprint of the hardcover 1st edition 1990

Reproduction of the figures: Gustav Dreher GmbH, Stuttgart

2121/3130-543210 - Printed on acid-free paper

To our families, with love

Authors and Contributors

HENRI-MARCEL HOOGEWOUD, M. D.
Department of Radiology, Cantonal Hospital
1700 Fribourg, Switzerland

GÜNTER RAGER, Prof. M. D.
Institute of Anatomy, University of Fribourg
1700 Fribourg, Switzerland

HANS-BEAT BURCH, M. D.
Department of Orthopedic Surgery, Cantonal Hospital
1700 Fribourg, Switzerland

PHILIPPE CERUTTI, M. D.
Department of Orthopedic Surgery
and Surgery of the Locomotive Apparatus
Division of Surgery, University Cantonal Hospital
1211 Geneva 4, Switzerland
Formerly: Department of Orthopedic Surgery
Cantonal Hospital
1700 Fribourg, Switzerland

GISELA RILLING, M. D.
Institute of Anatomy, University of Fribourg
1700 Fribourg, Switzerland

Acknowledgments

We would like to thank all those without whom this book could not have been written. We are especially indebted to the AO Stiftung ASIF Foundation, Bern, Switzerland for their generous assistance and their financial support. The technical skills of Alain Devaud, Christian Bähler, Samuel Grenier, and the other technicians of the Radiology Department of the Cantonal Hospital Fribourg were essential for the radiological part of this book. We are especially grateful to the anatomical preparators Robert Esseiva and Franz Jungo of the Institute of Anatomy, University of Fribourg for their great work in the preparation of the cadaver sections. The radiographic and photographic reproductions were made by Stuart Grainger and Jean-Luc Theytaz, and the illustrations were drawn by Pierre-François Bossy. We thank them very much for their work. We are also very grateful to Corinne Wicht, Elisabeth Simon, and Kathrin Haenni for their secretarial assistance.

Special thanks are due to Springer-Verlag, particularly to Ute Heilmann and David Roseveare, for their cooperation and assistance in the preparation of the book.

Our thanks also go to Philippe Cerutti for his contribution to the creation of the software supplied with the book and to Gisela Rilling and Egon Boedtker, whose thorough analysis of the anatomical sections was of great help.

Finally, thanks to our families for their patience and their love.

Fribourg, Spring 1990   H.-M. Hoogewoud
                                                          G. Rager
                                                          H.-B. Burch

Contents

Introduction

When computed tomography (CT) was developed and introduced by Hounsfield (1973), a new era of clinical diagnostic potential began. At the same time CT created new difficulties, in that the physicians who had to deal with the CT images were not acquainted with their interpretation. Therefore, it became necessary to compare CT scans with anatomical sections, which gave additional information by virtue of their higher resolution, the different colors and consistencies of the structures, and the possibility to trace these structures across several sections. Several atlases comparing CT scans and anatomical sections were published soon after the introduction of the new technique.

The resolving power of the new scanners has increased considerably, necessitating a renewed comparison between CT scans and anatomical sections. A threefold need for higher-quality anatomical sections has also become evident: First, tissue preservation should be excellent. Second, the sections should not be thicker than the scans obtained by the CT procedure. Third, the series of sections should be complete in order to permit three-dimensional reconstructions.

We have tried to meet these requirements, restricting ourselves to the analysis of the lower extremity. The leg had to be scanned serially, then cut in serial sections in such a way that the CT planes and the anatomical sections corresponded optimally. As many sections had to be illustrated as were necessary to demonstrate changes in the internal structure of the extremity wherever they occurred.

Since several books describing CT scanning techniques in detail have already been published, we were able to confine ourselves to three practical aspects. Our first aim was to provide an atlas for physicians, radiologists, and orthopedists which would serve as a quick reference for the interpretation of CT scans. Second, we wanted to describe the normal structure as a basis for the recognition of pathological processes. Our third aim was to provide an adequate means of teaching medical students cross-sectional anatomy, equipping them to use the new diagnostic techniques in their future clinical work.

In addition, we wanted to show how CT can be used for various measurements of the lower extremity. CT is also useful in assessing the coverage of the femoral head by the acetabulum, as explained in the

1

closing chapter and on the accompanying IBM-compatible floppy disk, which contains all the necessary software.

Fribourg, Spring 1990

Material and Techniques

CT Scanning

Requirements

High contrast between different structures is essential for good CT scans. Before embarking on this project we examined several fixed cadaver limbs and noted that fixation caused the CT number of fat and muscles to become similar and induced artifacts. We decided that for our purposes only a nonfixed cadaver could be used for CT scanning. The difficulty in identifying the individual muscles or groups of muscles if the tissue separating them has the same density as the muscle tissue itself is well known. As contrast enhancement could not be performed, we selected a relatively fatty cadaver in order to allow very fine structures, such as nerves, to be recognized on the CT scans and to be identified by reference to the anatomical sections. On several occasions, however, structures seen on the anatomical sections could not be visualized on the CT scans. The above-mentioned advantage of studying a fatty cadaver is accompanied by the disadvantage of a greater chance of encountering structural changes due to osteoarthritis. Unfortunately, several such alterations were observed in the hip and knee regions.

Examination Technique

One day after death the cadaver of a 69-year-old woman was examined in a supine position from the crista iliaca to the foot using a Somatom DRH CT scanner (Siemens, Erlangen, FRG). The scan thickness was 2 mm throughout. We systematically used a 512×512 matrix for data acquisition. The milliampere/second (mAS) settings were put at the highest level, and the machine was recalibrated frequently in order to avoid artifacts as much as possible. The window settings we chose represented a compromise among the requirements for printing, for contrast between soft tissues, and for visualization of bony structures.

The table feed was varied: 2 mm in the hip, knee, and foot regions, 10 mm in the thigh and lower leg regions. The scan levels were marked on the skin around the whole circumference of the limb with waterproof ink. The foot was in a supinated extended position with an outward rotation of 15° due to cadaveric rigidity, with the result that the foot region could not be scanned totally using the standard planes, parallel and perpendicular to the planta pedis. To illustrate the anatomy of the foot, we took scans of the right foot of one of the authors. For this reason, there are no corresponding anatomical sections in the foot region. For the reader's convenience, the level of each scan and section is indicated on an accompanying drawing.

We have included some CT arthrograms of the hip, the knee, and the ankle. These should not only show the structure of the articular cavities but also demonstrate the technical potential of modern scanners.

Anatomical Sections

Immediately after the scanning procedure the cadaver was fixed by injection of Jores I solution (500 ml sal carol. factit., 500 ml formaldehyde technic., 500 ml chloral hydrate technic., aq. dest. ad 10 l) into the brachial and femoral arteries. After a few days of fixation it was placed into a refrigerator at −20 °C. When the cadaver was deeply frozen, the extremity which had been examined in the CT scanner was isolated. In order to stabilize the longitudinal axis and to guarantee close correspondence of the section planes to the planes used in the scanner, the limb was embedded in a highly concentrated gelatin solution which was hardened in rectangular molds manufactured specially for this experiment. For serial sections an industrial band saw was adapted in such a way that sections could be cut at a minimal thickness of 2 mm. The saw-blade was cooled with liquid nitrogen. Thus, the loss could be minimized to approximately 0.5 mm per section.

Photography

Each section was photographed from both sides with both color slide film and black-and-white print film. The color slides were used only to identify structures which could not be clearly distinguished in black-and-white enlargements. In some cases the original sections had to be studied again for identification of these structures. The final enlargement was chosen in such a way that the calibration had to be changed only once, namely in the knee region (no. 24 in the series).

The most important anatomical sections were compared with the corresponding CT scans. Usually the structures described could be identified in both the section and the scan and are of clinical relevance. In some cases which seemed to us particularly important, the description of the anatomical sections is more detailed than that of the corresponding CT scans.

In several cases we consulted the following textbooks for identification of certain structures and for nomenclature: Hafferl (1969); Lang and Wachsmuth (1972); Gambarelli et al. (1977); Peterson (1980); Benninghoff (1985); Rauber/Kopsch (1988).

Atlas

The denomination of anatomical structures in the following illustrations strictly follows the fifth edition of *Nomina Anatomica* (1983), the only nomenclature accepted worldwide. The following abbreviations are used throughout:

A., Aa.	Arteria, arteriae
V., Vv.	Vena, venae
N., Nn.	Nervus, nervi
Lig., Ligg.	Ligamentum, ligamenta
M., Mm.	Musculus, musculi

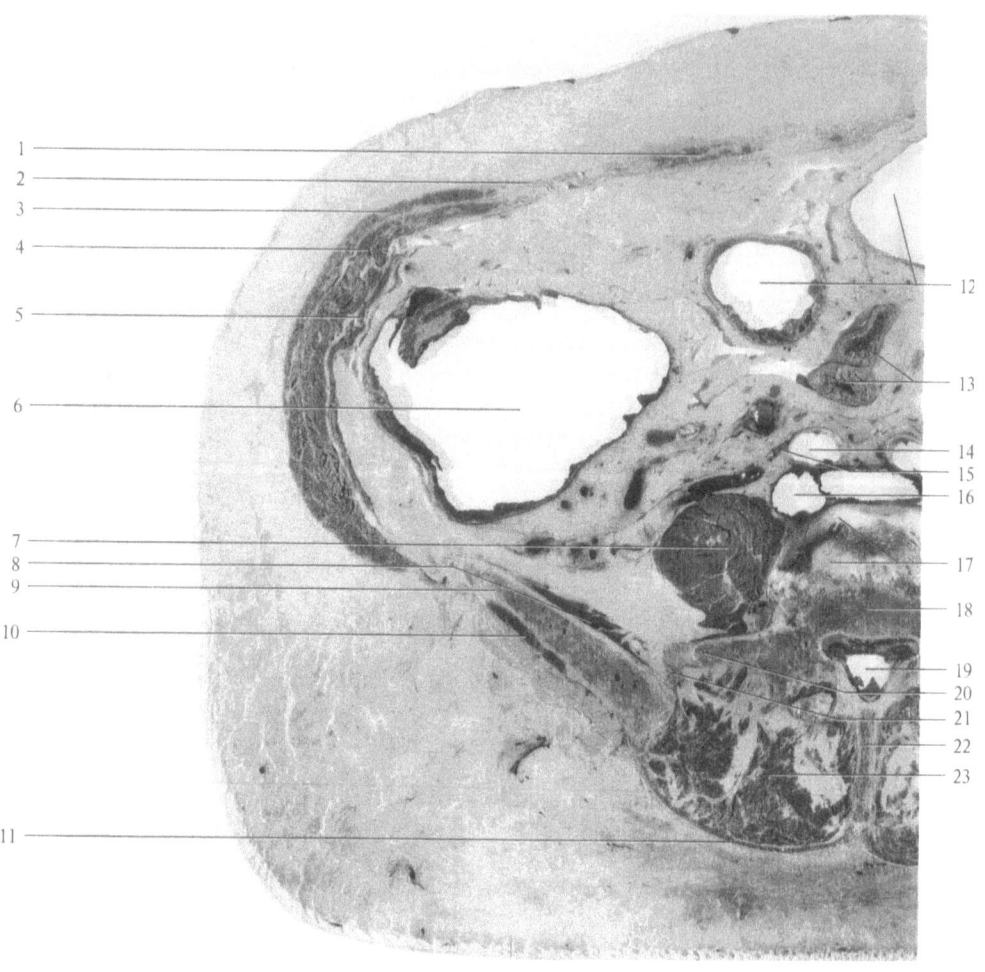

1
2
3
4
5
6
7
8
9
10
11

12
13
14
15
16
17
18
19
20
21
22
23

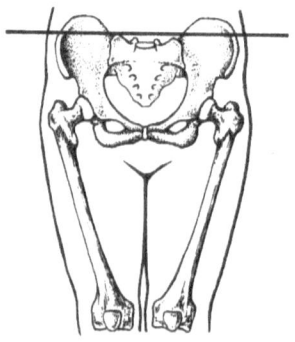

Plate 1

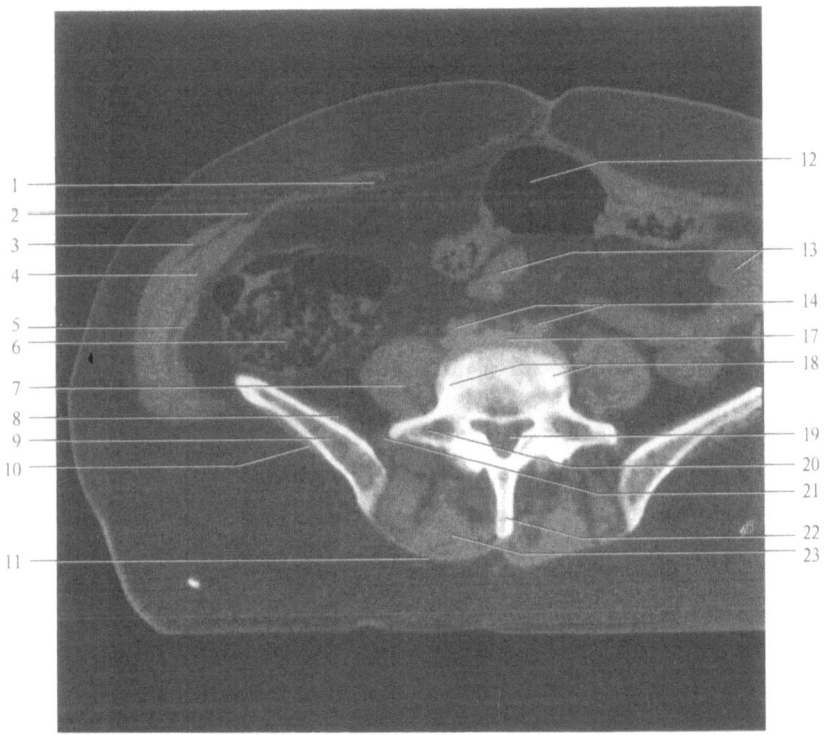

1 M. rectus abdominis; 2 Aponeurosis m. obliqui externi abdominis; 3 M. obliquus externus abdominis; 4 M. obliquus internus abdominis; 5 M. transversus abdominis; 6 Caecum; 7 M. psoas major; 8 M. iliacus; 9 Ala ossis ilii; 10 M. gluteus medius; 11 Fascia thoracolumbalis; 12 Colon transversum; 13 Ileum; 14 A. iliaca communis (CT: Vasa iliaca communia); 15 Ureter; 16 V. iliaca communis; 17 Discus intervertebralis L4–L5; 18 Corpus vertebrae L5; 19 Canalis vertebralis; 20 Processus costalis; 21 Lig. iliolumbale; 22 Processus spinosus; 23 M. erector spinae

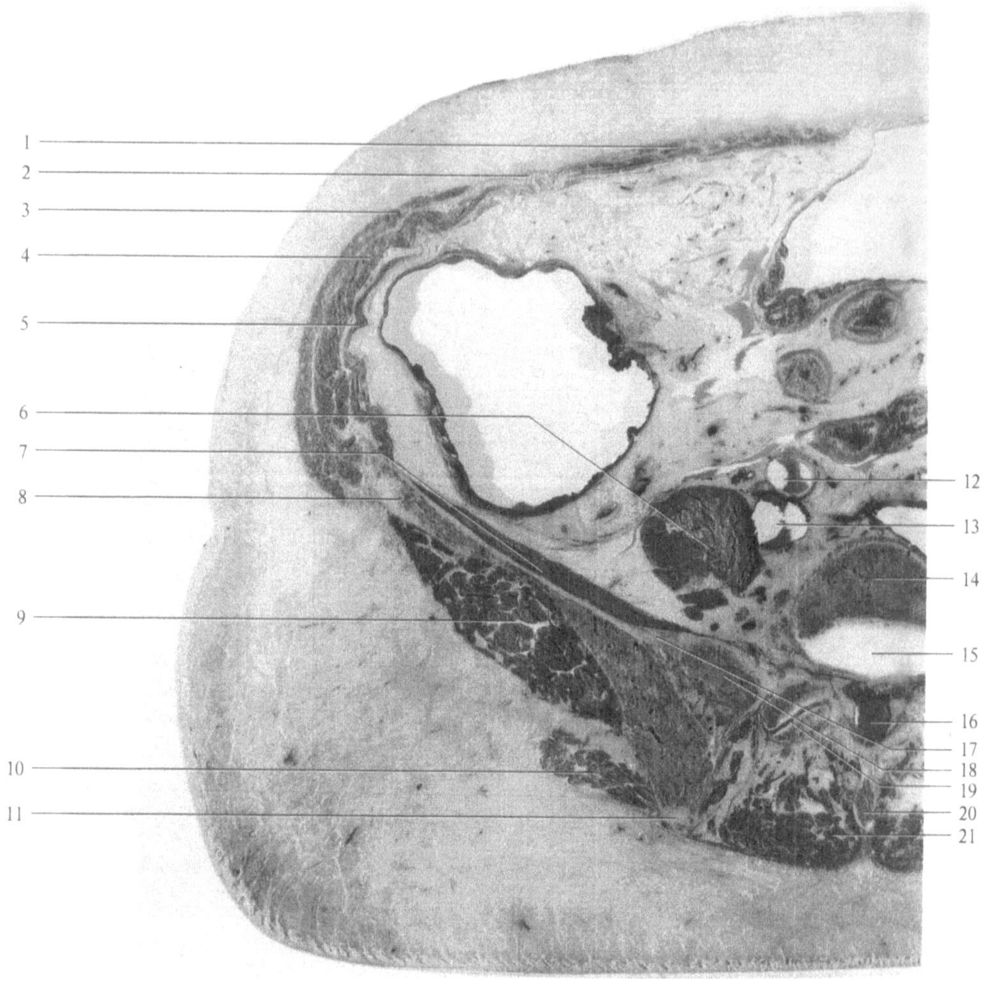

1
2
3
4
5
6
7
8
9
10
11
12
13
14
15
16
17
18
19
20
21

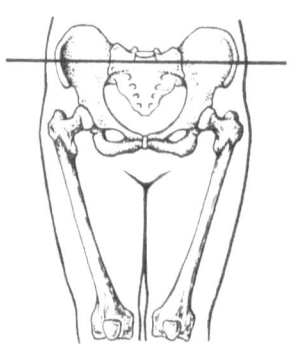

Plate 2

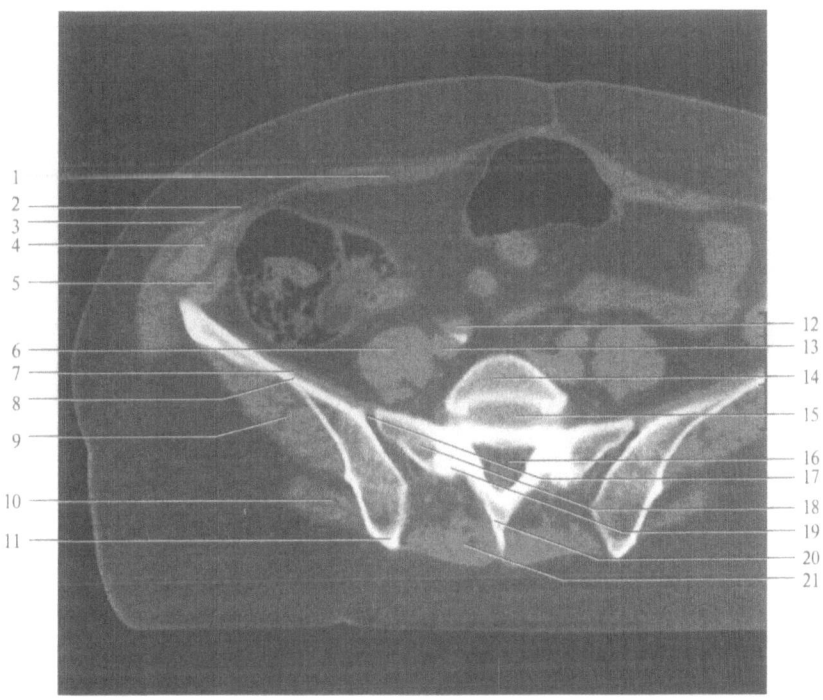

1 M. rectus abdominis; *2* Aponeurosis m. obliqui externi abdominis; *3* M. obliquus externus abdominis; *4* M. obliquus internus abdominis; *5* M. transversus abdominis; *6* M. psoas major; *7* M. iliacus; *8* Ala ossis ilii; *9* M. gluteus medius; *10* M. gluteus maximus; *11* Spina iliaca posterior superior; *12* A. iliaca communis; *13* V. iliaca communis; *14* Corpus vertebrae L5; *15* Discus intervertebralis L5–S1; *16* Canalis vertebralis; *17* Ligg. sacro-iliaca anteriora (CT: Articulatio sacro-iliaca); *18* Ala sacralis; *19* Articulatio lumbosacralis; *20* Processus spinosus L5; *21* M. erector spinae

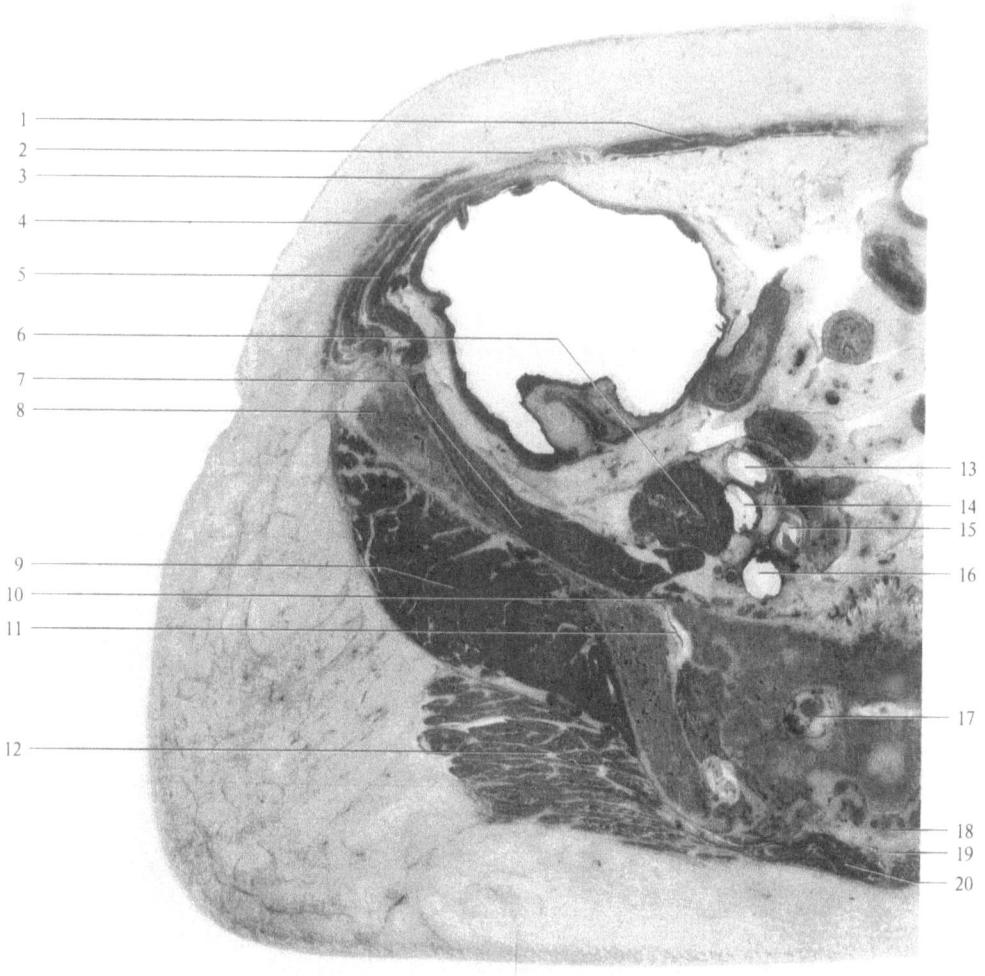

1
2
3
4
5
6
7
8

13
14
15
16

9
10
11

17

12

18
19
20

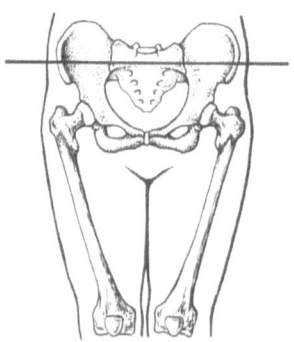

12

Plate 3

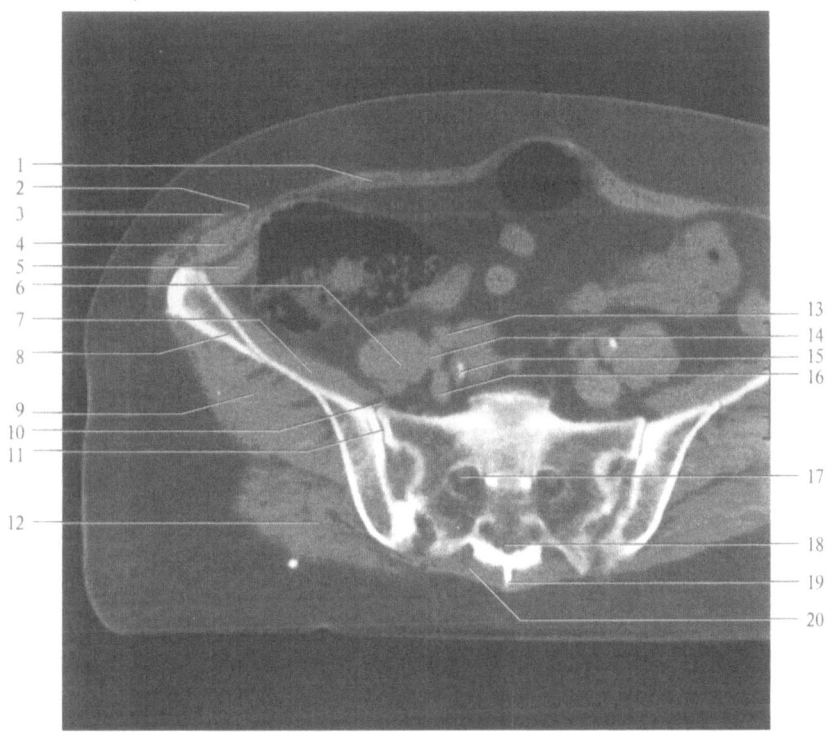

1 M. rectus abdominis; 2 Aponeurosis m. obliqui externi abdominis; 3 M. obliquus externus abdominis; 4 M. obliquus internus abdominis; 5 M. transversus abdominis; 6 M. psoas major; 7 M. iliacus; 8 Ala ossis ilii; 9 M. gluteus medius; 10 Ligg. sacro-iliaca anteriora; 11 Articulatio sacro-iliaca; 12 M. gluteus maximus; 13 A. iliaca externa; 14 V. iliaca externa; 15 A. iliaca interna; 16 V. iliaca interna; 17 Foramen sacrale; 18 Canalis sacralis; 19 Crista sacralis mediana; 20 M. erector spinae

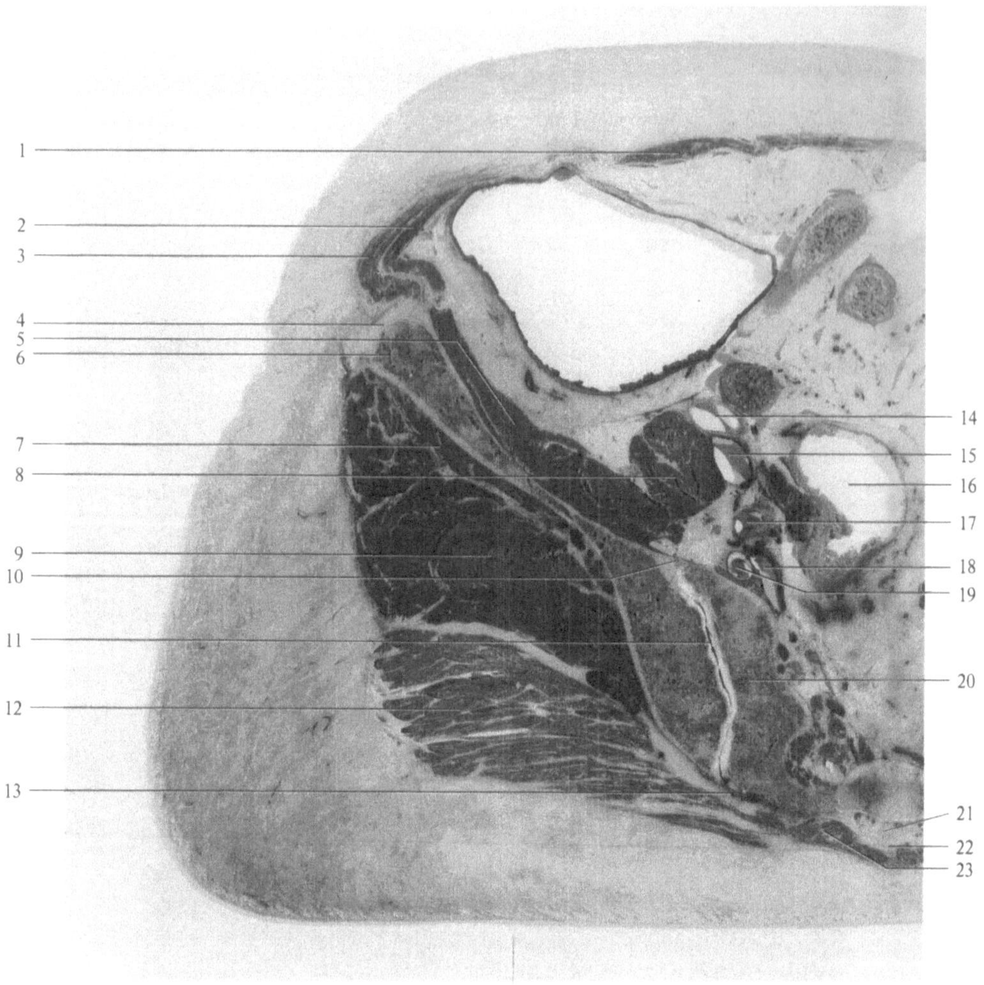

1

2
3

4
5
6

7
8

9
10

11

12

13

14

15
16
17

18
19

20

21
22
23

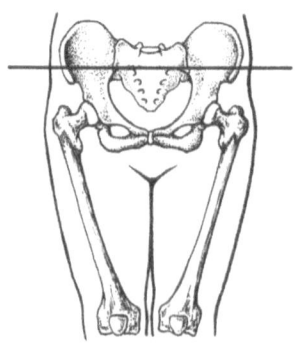

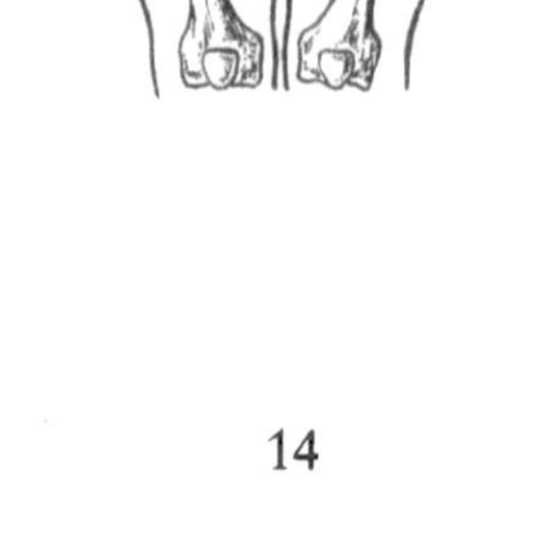

14

Plate 4

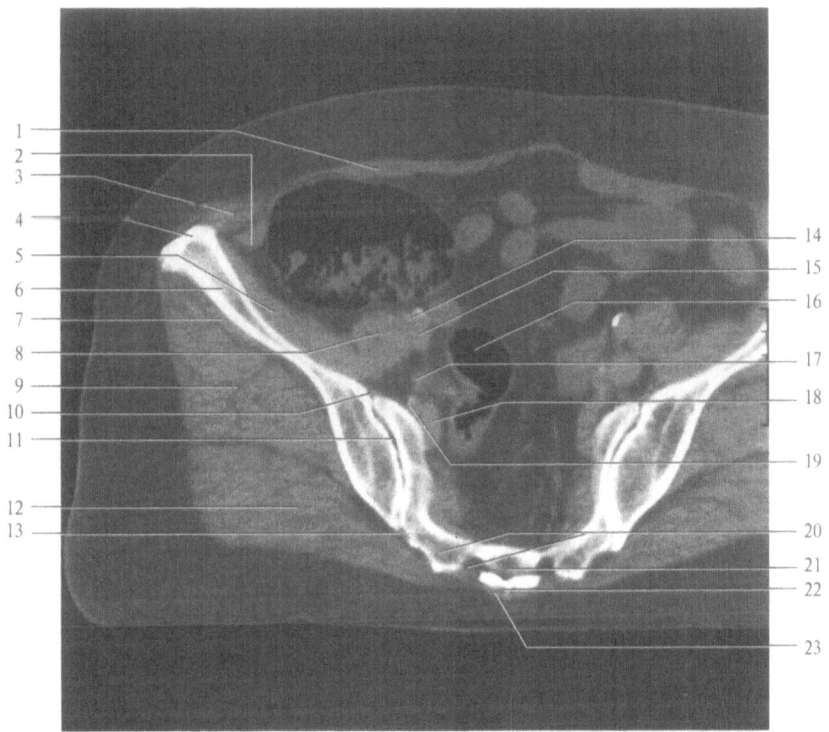

1 M. rectus abdominis; 2 M. transversus abdominis; 3 M. obliquus internus abdominis; 4 Crista iliaca; 5 M. iliacus; 6 Ala ossis ilii; 7 M. gluteus minimus; 8 M. psoas major; 9 M. gluteus medius; 10 Ligg. sacro-iliaca anteriora; 11 Articulatio sacro-iliaca; 12 M. gluteus maximus; 13 Ligg. sacro-iliaca posteriora; 14 A. iliaca externa; 15 V. iliaca externa; 16 Colon sigmoideum; 17 A. iliaca interna; 18 V. iliaca interna; 19 A. glutea superior; 20 Os sacrum (CT: Os sacrum, Foramen sacrale); 21 Canalis sacralis; 22 Crista sacralis mediana; 23 M. erector spinae

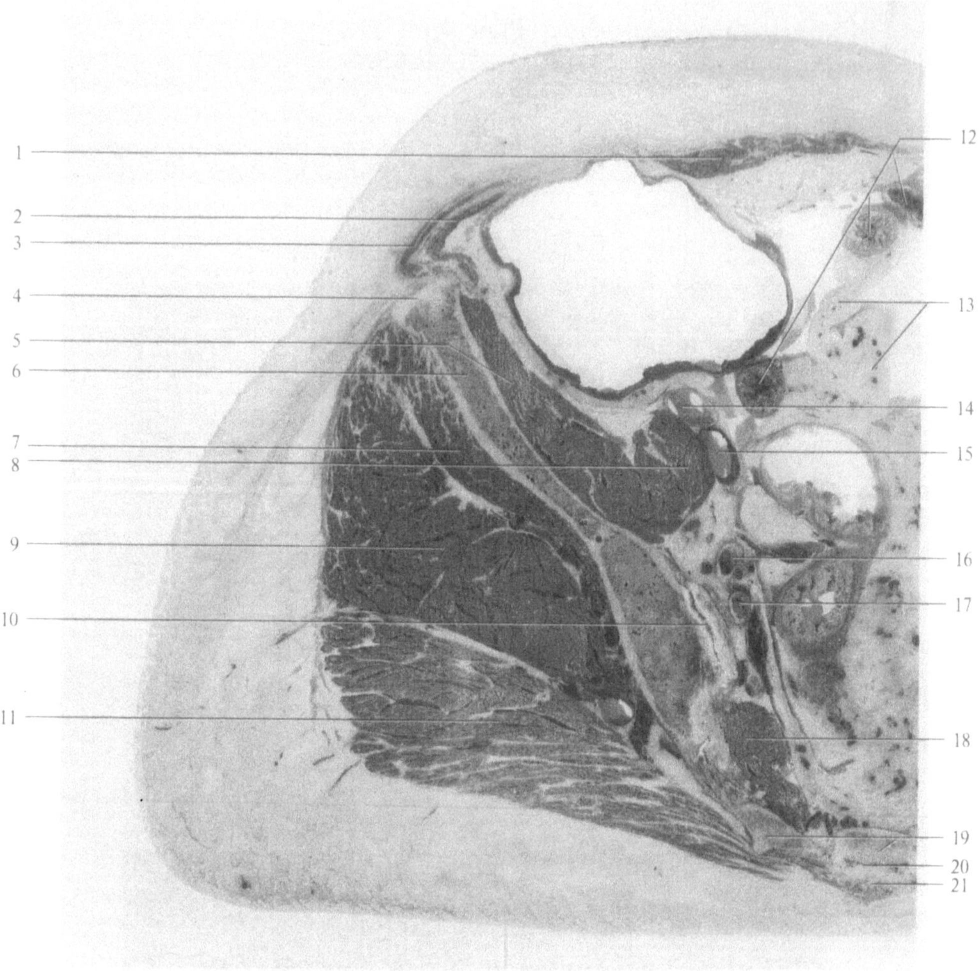

1

2
3

4

5
6

7
8

9

10

11

12

13

14

15

16

17

18

19
20
21

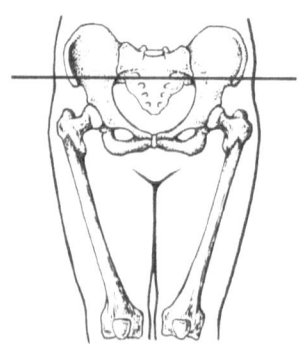

Plate 5

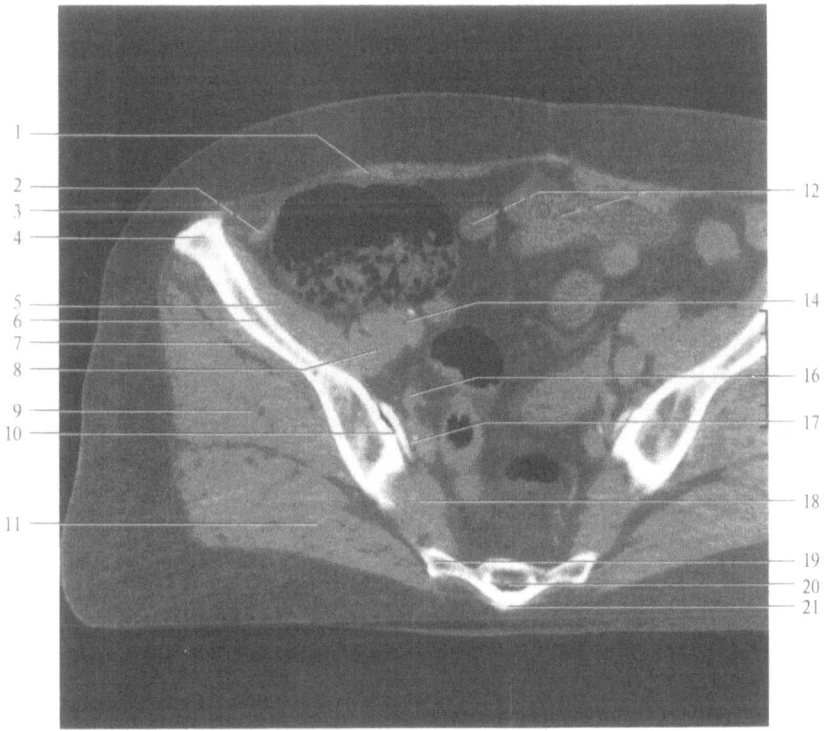

1 M. rectus abdominis; 2 M. transversus abdominis; 3 M. obliquus internus abdominis; 4 Crista iliaca; 5 M. iliacus; 6 Ala ossis ilii; 7 M. gluteus minimus; 8 M. psoas major; 9 M. gluteus medius; 10 Articulatio sacro-iliaca; 11 M. gluteus maximus; 12 Ileum; 13 Mesenterium; 14 A. iliaca externa; 15 V. iliaca externa; 16 A. iliaca interna; 17 A. glutea superior; 18 M. piriformis; 19 Os sacrum; 20 Canalis sacralis; 21 Crista sacralis mediana

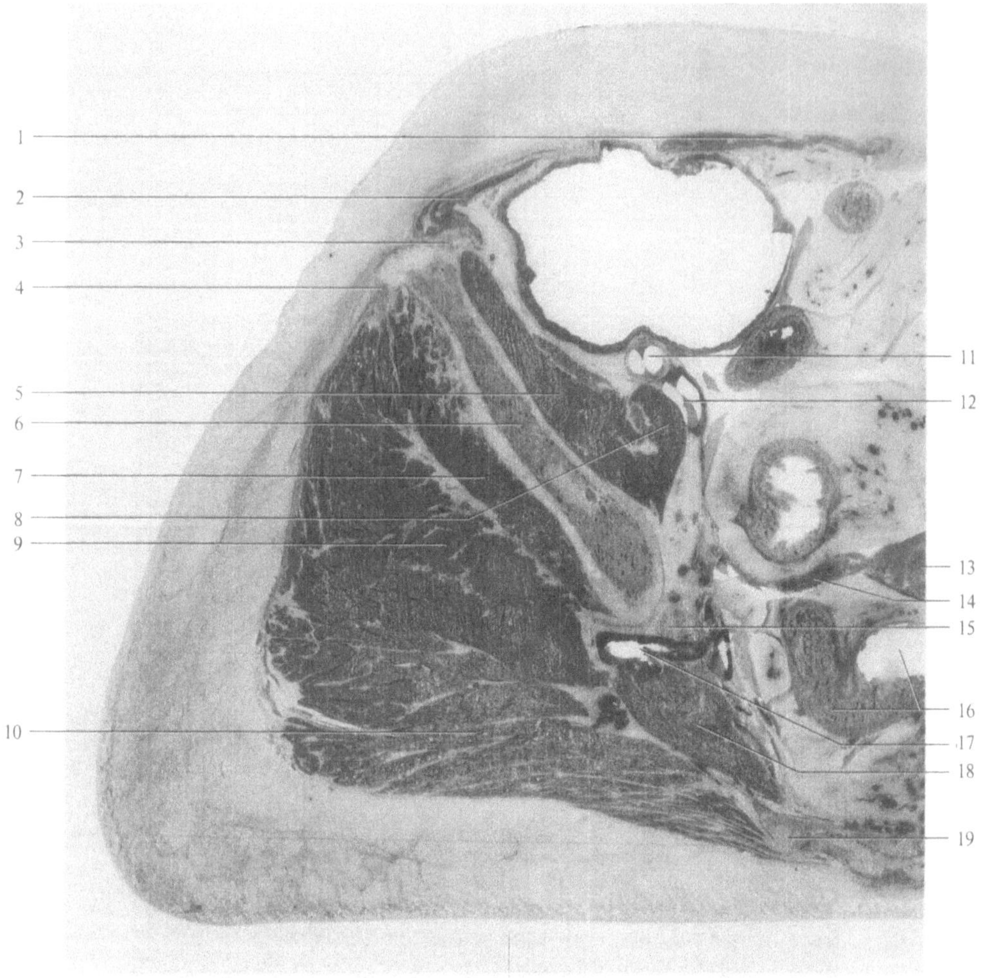

1

2

3

4

5

6

7

8

9

10

11

12

13

14

15

16

17

18

19

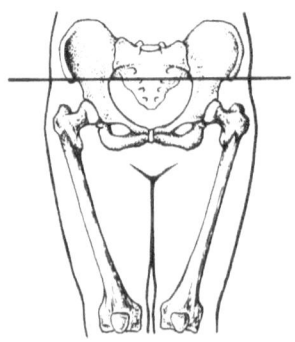

Plate 6

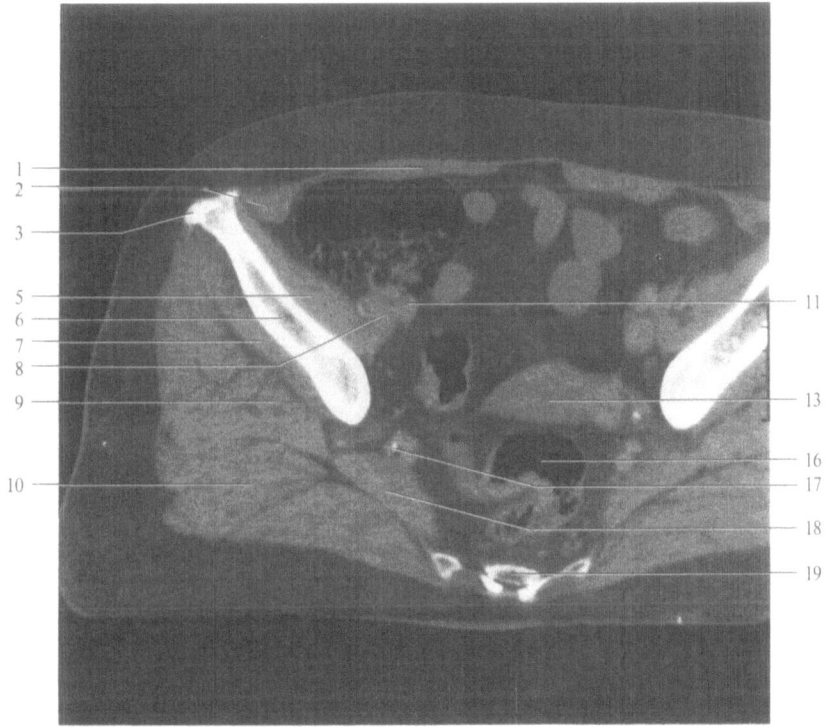

1 M. rectus abdominis; *2* Mm. transversus et obliquus internus abdominis; *3* Spina iliaca anterior superior; *4* Origo m. tensoris fasciae latae; *5* M. iliacus; *6* Ala ossis ilii; *7* M. gluteus minimus; *8* M. psoas major; *9* M. gluteus medius; *10* M. gluteus maximus; *11* A. iliaca externa; *12* V. iliaca externa; *13* Uterus; *14* Tuba uterina; *15* N. ischiadicus; *16* Rectum; *17* Vasa glutea superiora; *18* M. piriformis; *19* Os sacrum

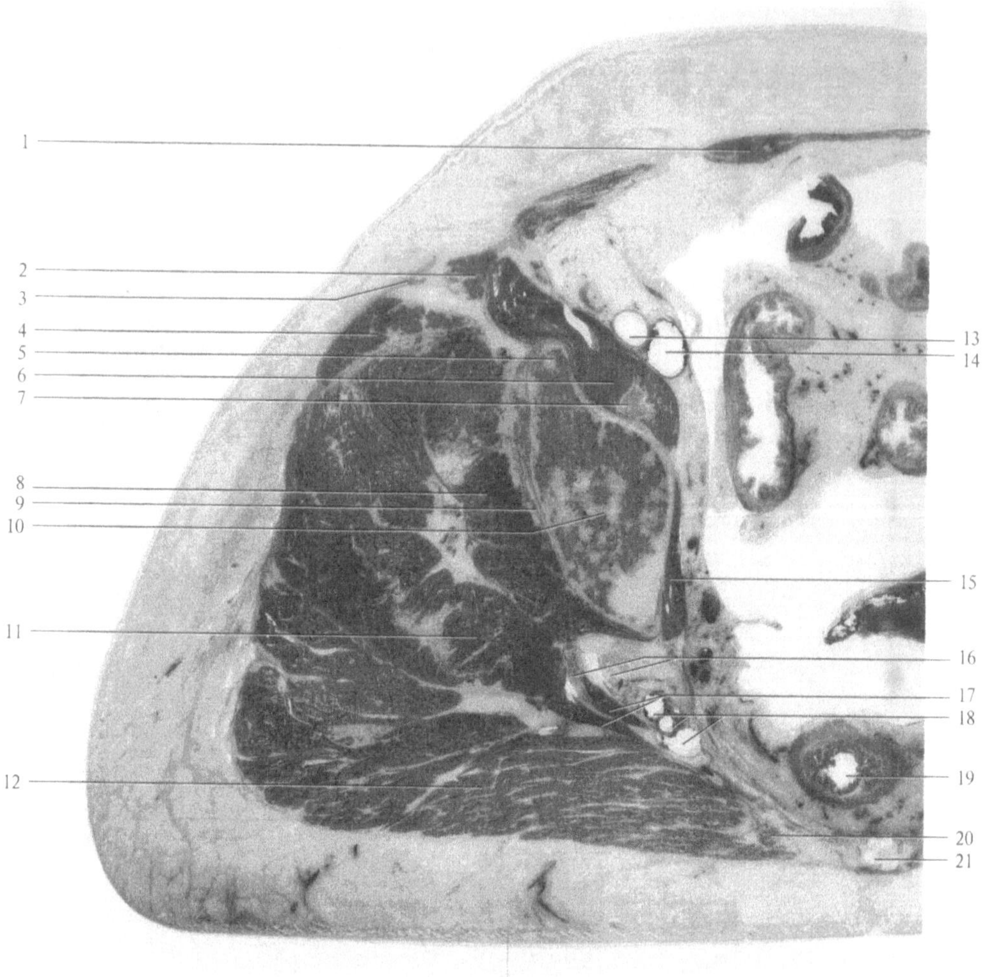

1

2
3
4
5
6
7

8
9
10

11

12

13
14

15

16
17
18

19

20
21

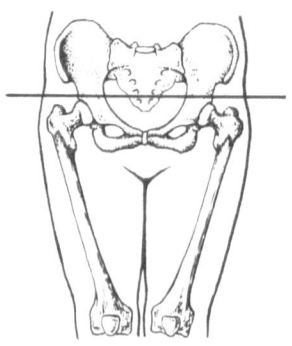

Plate 7

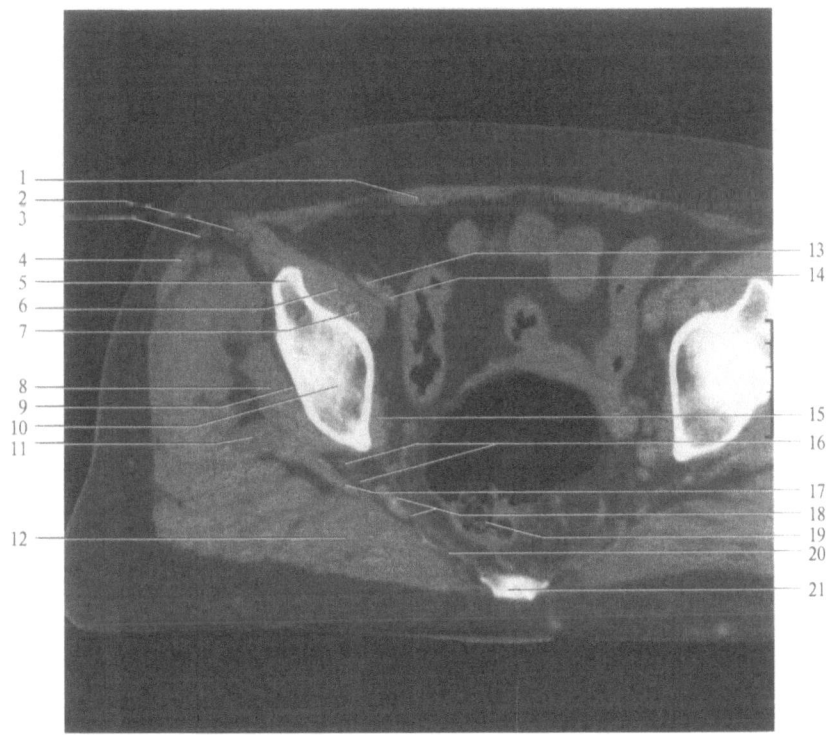

1 M. rectus abdominis; 2 M. sartorius; 3 Fascia lata; 4 M. tensor fasciae latae; 5 Spina iliaca anterior inferior; 6 M. iliopsoas; 7 Tendo m. psoatis majoris; 8 M. gluteus minimus; 9 Tendo capitis reflexi m. recti femoris; 10 Corpus ossis ilii; 11 M. gluteus medius; 12 M. gluteus maximus; 13 A. femoralis; 14 V. femoralis; 15 M. obturator internus; 16 N. ischiadicus; 17 M. piriformis; 18 Vasa glutea inferiora; 19 Rectum; 20 Ligg. sacrotuberale et sacrospinale; 21 Os coccygis

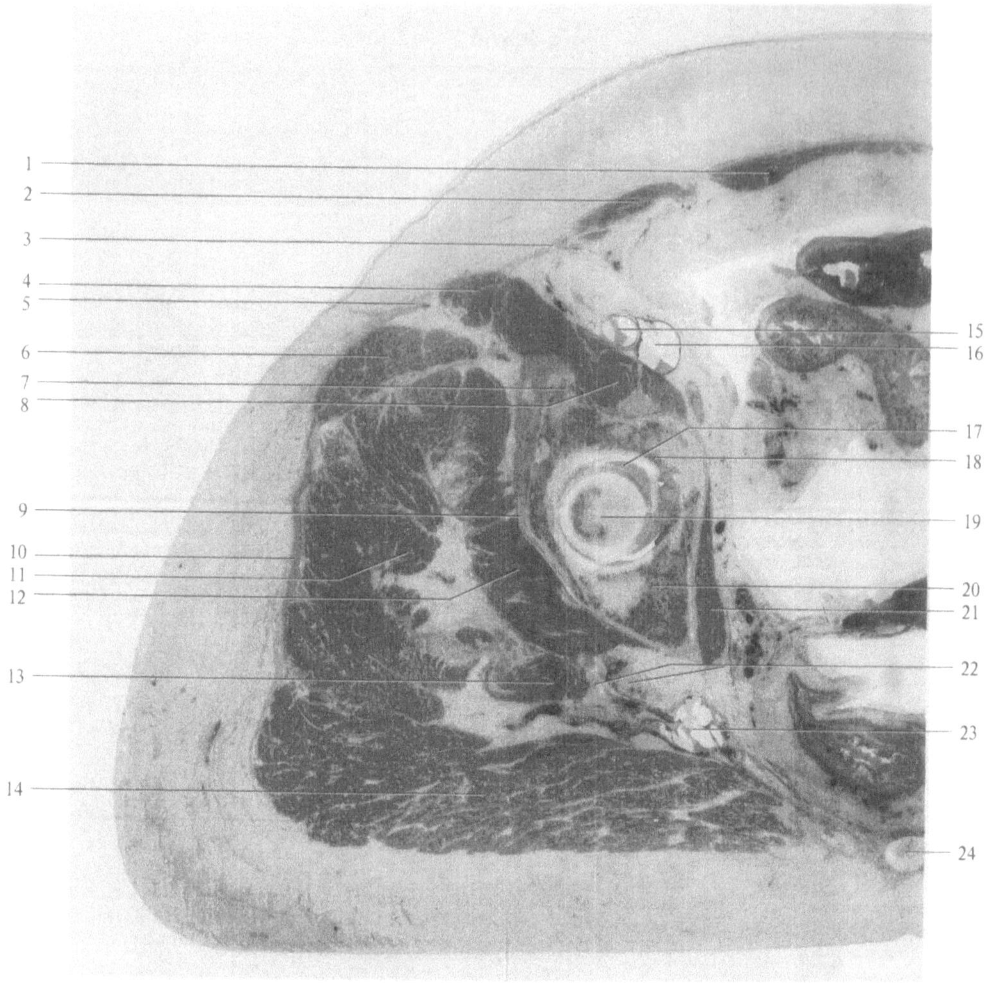

1

2

3

4
5

6

7
8

9

10
11
12

13

14

15
16

17
18

19

20
21

22

23

24

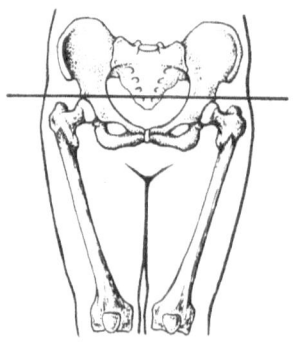

22

Plate 8

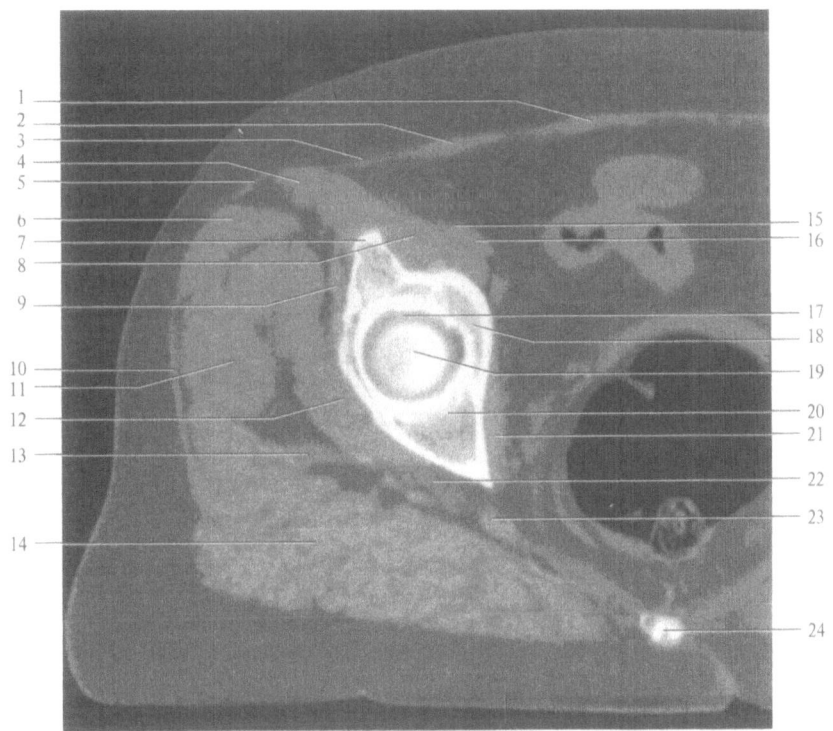

1 M. rectus abdominis; 2 Mm. obliquus internus et transversus abdominis; 3 Lig. inguinale; 4 M. sartorius; 5 Fascia lata; 6 M. tensor fasciae latae; 7 Spina iliaca anterior inferior; 8 M. iliopsoas; 9 Tendo capitis reflexi m. recti femoris; 10 Tractus iliotibialis; 11 M. gluteus medius; 12 M. gluteus minimus; 13 M. piriformis; 14 M. gluteus maximus; 15 A. femoralis; 16 V. femoralis; 17 Articulatio coxae; 18 Acetabulum; 19 Caput ossis femoris; 20 Corpus ossis ilii; 21 M. obturator internus; 22 N. ischiadicus; 23 Vasa glutea inferiora; 24 Os coccygis

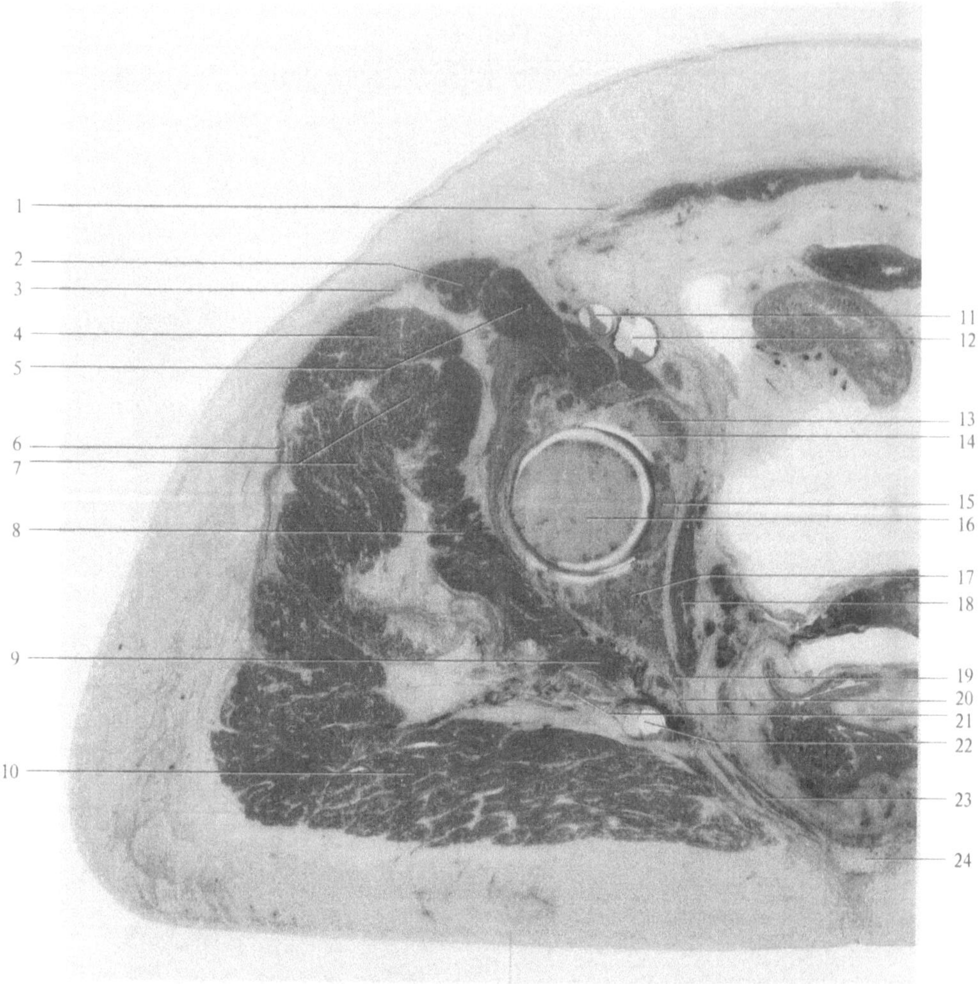

1
2
3
4
5

6
7

8

9

10

11
12

13
14

15
16

17
18

19
20
21
22

23

24

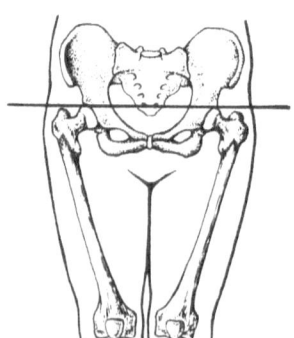

Plate 9

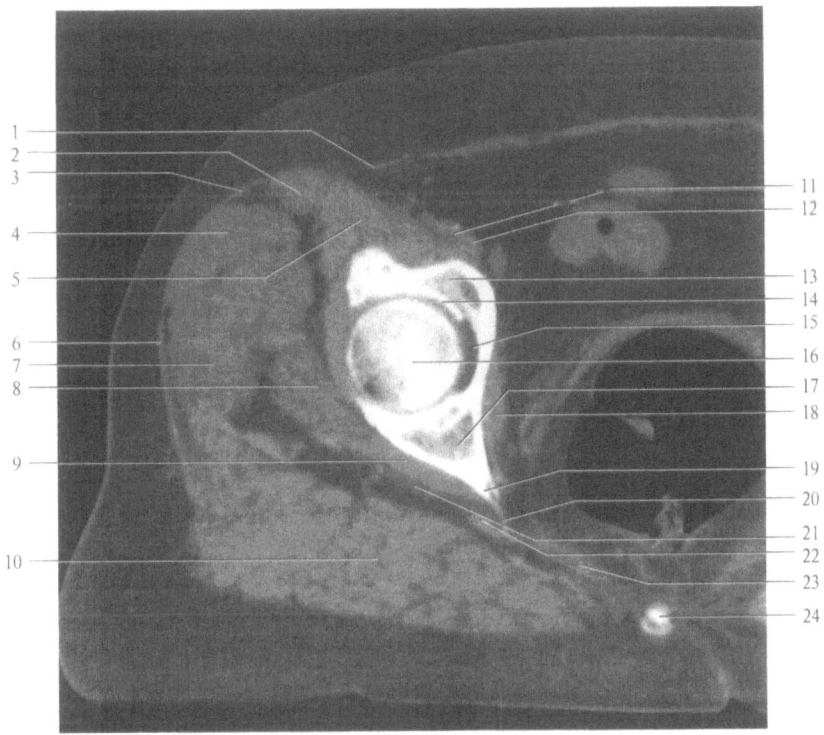

1 Lig. inguinale; 2 M. sartorius; 3 Fascia lata; 4 M. tensor fasciae latae; 5 M. iliopsoas; 6 Tractus iliotibialis; 7 M. gluteus medius; 8 M. gluteus minimus; 9 M. gemellus
superior; 10 M. gluteus maximus; 11 A. femoralis; 12 V. femoralis; 13 Corpus ossis
pubis; 14 Facies lunata; 15 Fossa acetabuli et Lig. capitis femoris; 16 Caput
ossis femoris; 17 Corpus ossis ischii; 18 M. obturator internus; 19 Spina ischiadica;
20 Lig. sacrospinale; 21 N. ischiadicus; 22 Vasa glutea inferiora; 23 M. levator ani;
24 Os coccygis

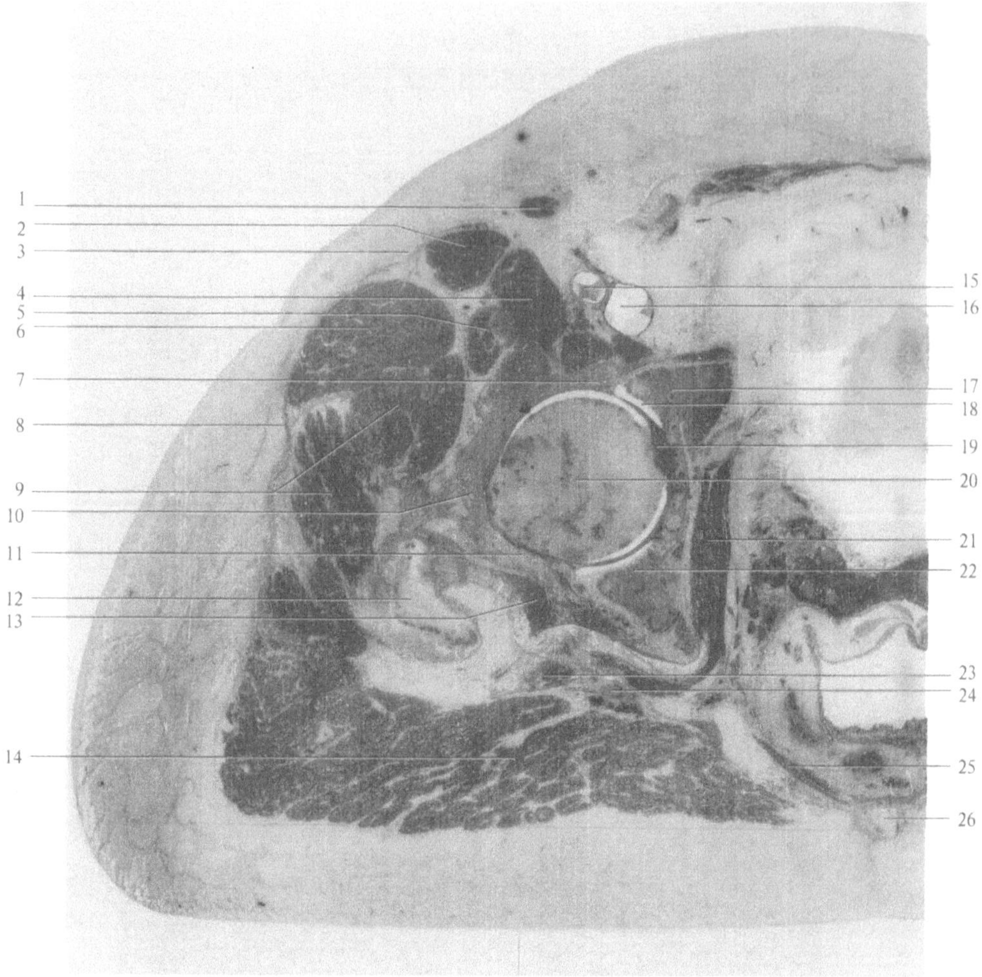

1

2

3

4

5

6

7

8

9

10

11

12

13

14

15

16

17

18

19

20

21

22

23

24

25

26

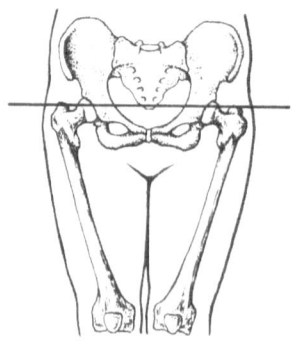

26

Plate 10

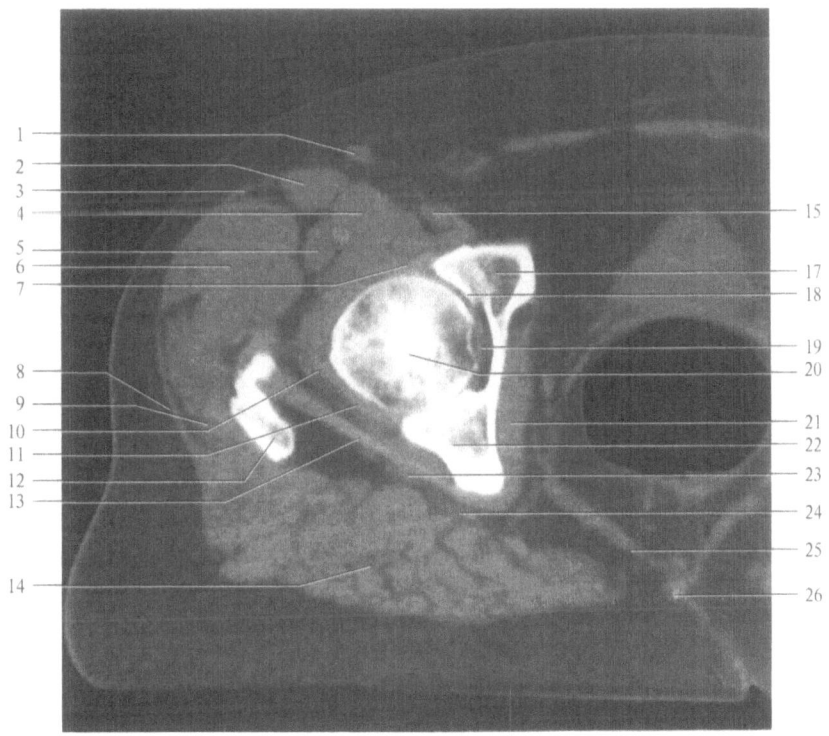

1 Nodus lymphaticus inguinalis; *2* M. sartorius; *3* Fascia lata; *4* M. iliopsoas; *5* M. rectus femoris; *6* M. tensor fasciae latae; *7* Lig. pubofemorale; *8* Tractus iliotibialis; *9* M. gluteus medius; *10* Capsula articularis; *11* Lig. ischiofemorale; *12* Trochanter major; *13* M. gemellus superior; *14* M. gluteus maximus; *15* A. femoralis; *16* V. femoralis; *17* Corpus ossis pubis; *18* Facies lunata; *19* Fossa acetabuli et Lig. capitis femoris; *20* Caput ossis femoris; *21* M. obturator internus; *22* Corpus ossis ischii; *23* N. ischiadicus; *24* Vasa glutea inferiora; *25* M. levator ani; *26* Os coccygis

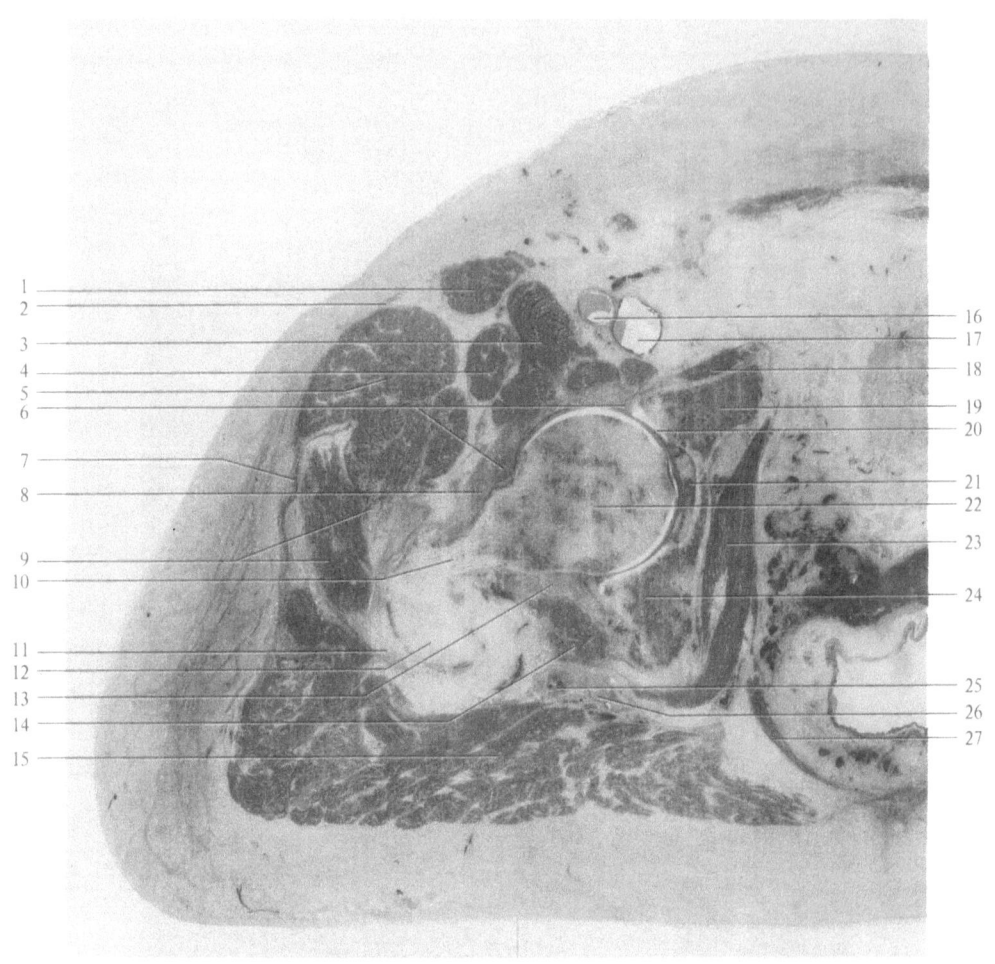

1
2
3
4
5
6

7
8

9
10

11
12
13
14
15

16
17
18
19
20

21
22
23

24

25
26
27

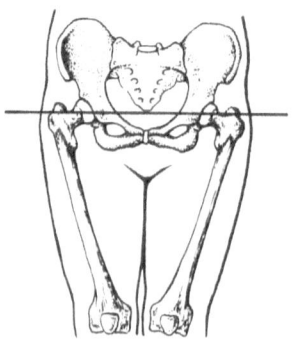

Plate 11

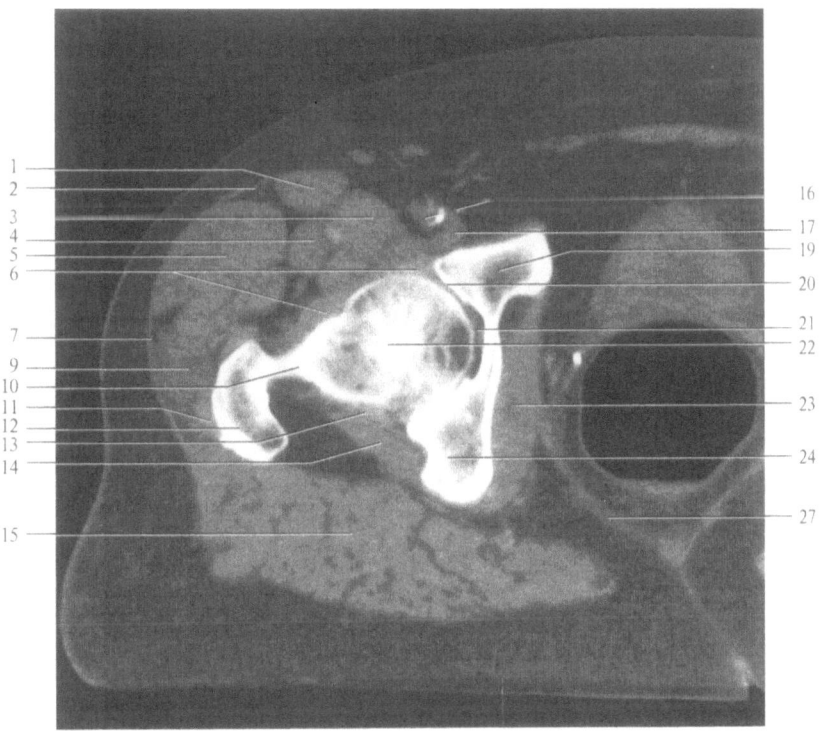

1 M. sartorius; *2* Fascia lata; *3* M. iliopsoas; *4* M. rectus femoris; *5* M. tensor fasciae latae; *6* Lig. pubofemorale, Labrum acetabulare; *7* Tractus iliotibialis; *8* Capsula articularis; *9* M. gluteus medius; *10* Collum ossis femoris; *11* Tendo m. glutei medii; *12* Trochanter major; *13* Lig. ischiofemorale; *14* M. quadratus femoris; *15* M. gluteus maximus; *16* A. fcmoralis; *17* V. femoralis; *18* M. pectineus; *19* Ramus superior ossis pubis; *20* Facies lunata; *21* Fossa acetabuli et Lig. capitis femoris; *22* Caput ossis femoris; *23* M. obturator internus; *24* Corpus ossis ischii; *25* N. ischiadicus; *26* Vasa glutea inferiora; *27* M. levator ani

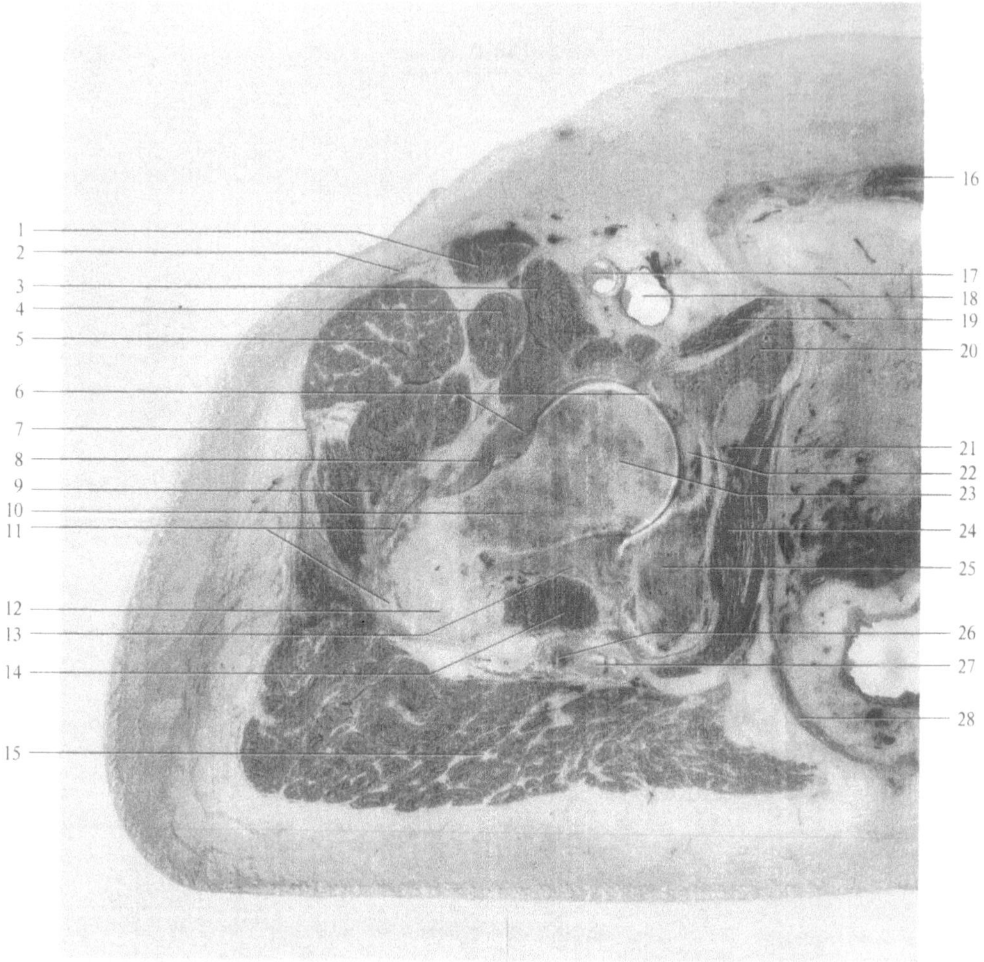

1
2
3
4
5
6
7
8
9
10
11
12
13
14
15

16
17
18
19
20

21
22
23
24
25

26
27
28

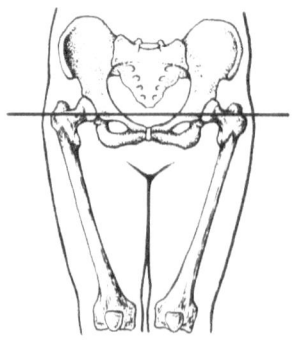

Plate 12

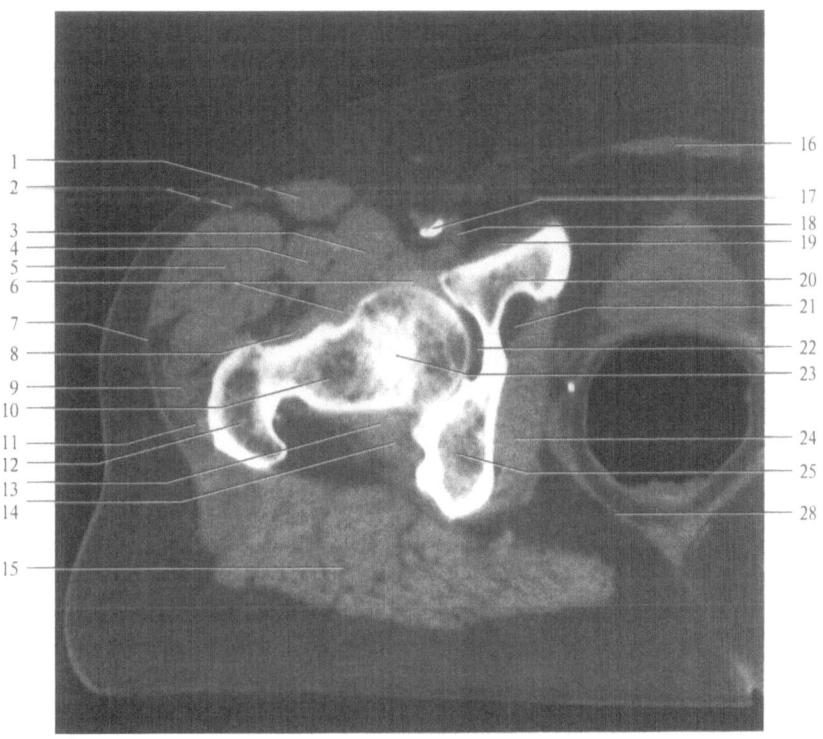

1 M. sartorius; *2* Fascia lata; *3* M. iliopsoas; *4* M. rectus femoris; *5* M. tensor fasciae latae; *6* Lig. pubofemorale, Labrum acetabulare; *7* Tractus iliotibialis; *8* Capsula articularis; *9* M. gluteus medius; *10* Collum ossis femoris; *11* Tendo m. glutei medii; *12* Trochanter major; *13* Lig. ischiofemorale; *14* M. quadratus femoris; *15* M. gluteus maximus; *16* M. rectus abdominis; *17* A. femoralis; *18* V. femoralis; *19* M. pectineus; *20* Ramus superior ossis pubis; *21* Canalis obturatorius; *22* Fossa acetabuli et Lig. capitis femoris; *23* Caput ossis femoris; *24* M. obturator internus; *25* Corpus ossis ischii; *26* N. ischiadicus; *27* Vasa glutea inferiora; *28* M. levator ani

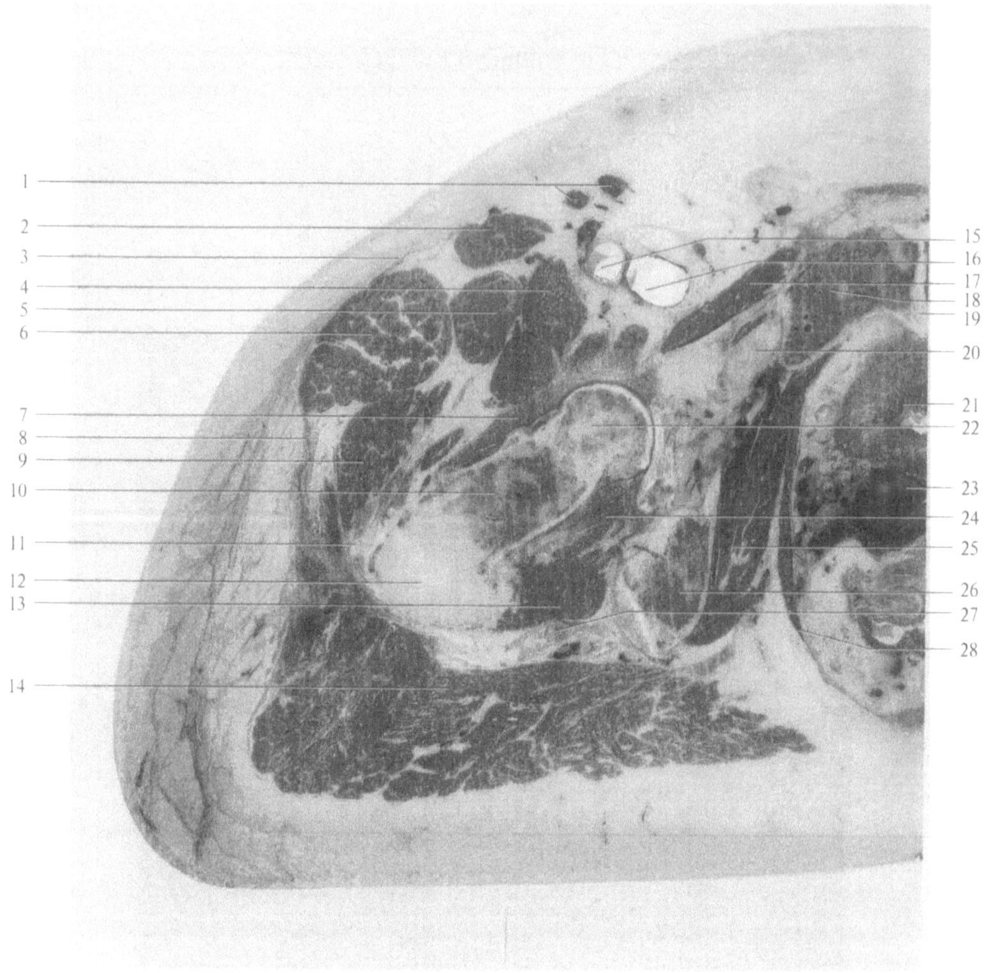

1

2
3
4
5
6

7
8
9
10

11

12
13

14

15
16
17
18
19

20

21
22

23

24
25

26
27
28

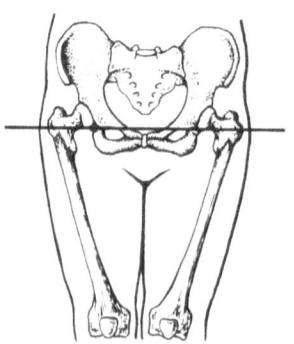

Plate 13

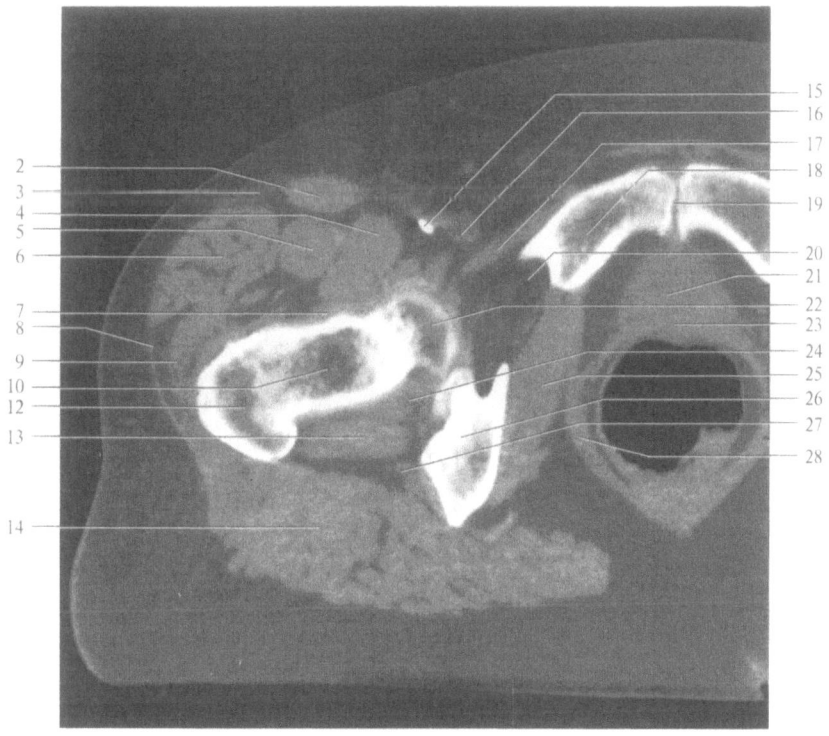

1 Nodi lymphatici inguinales; *2* M. sartorius; *3* Fascia lata; *4* M. iliopsoas; *5* M. rectus femoris; *6* M. tensor fasciae latae; *7* Capsula articularis, Ligg. articularia; *8* Tractus iliotibialis; *9* M. gluteus medius; *10* Collum ossis femoris; *11* Tendo m. glutei medii; *12* Trochanter major; *13* M. quadratus femoris; *14* M. gluteus maximus; *15* A. femoralis; *16* V. femoralis; *17* M. pectineus; *18* Os pubis; *19* Symphysis pubica; *20* Foramen obturatum, A. obturatoria, V. obturatoria, N. obturatorius; *21* Vesica urinaria; *22* Caput ossis femoris; *23* Cervix uteri; *24* M. obturator externus; *25* M. obturator internus; *26* Corpus ossis ischii; *27* N. ischiadicus; *28* M. levator ani

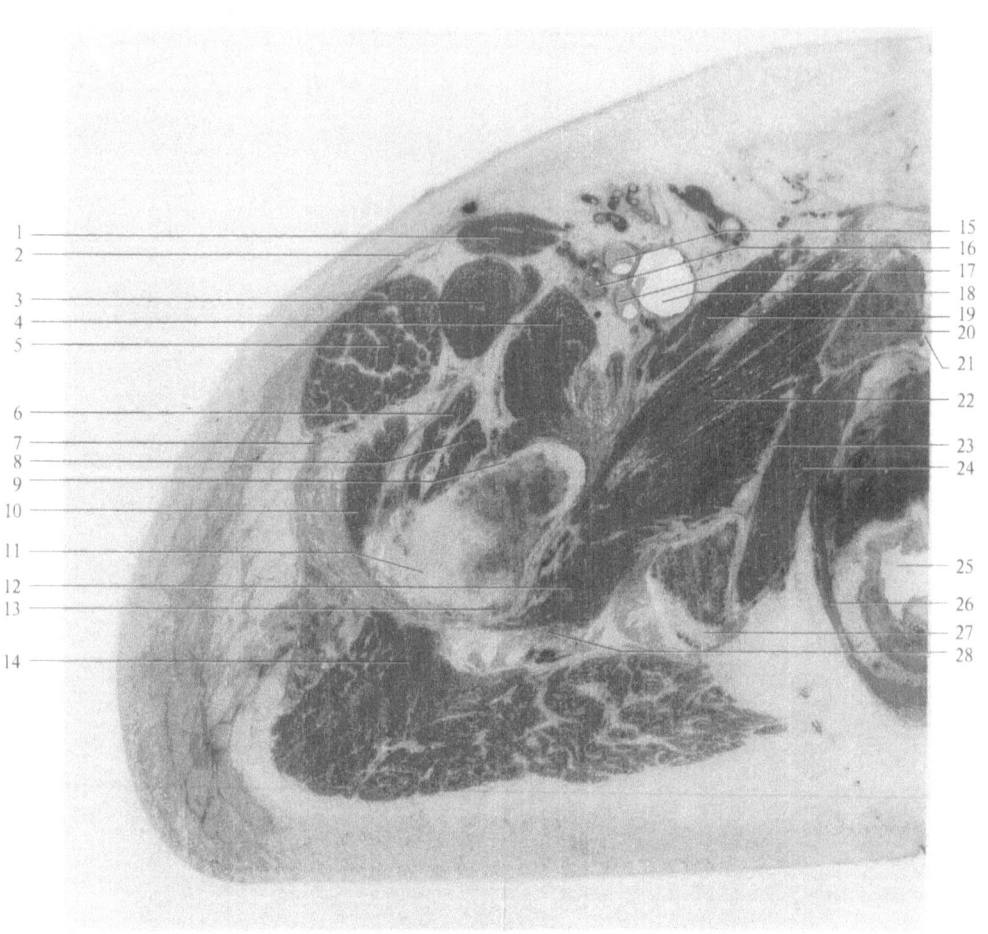

1
2

3
4
5

6
7
8
9
10

11
12
13

14

15
16
17
18
19
20
21

22

23
24

25

26
27
28

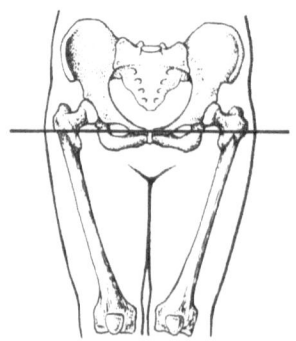

34

Plate 14

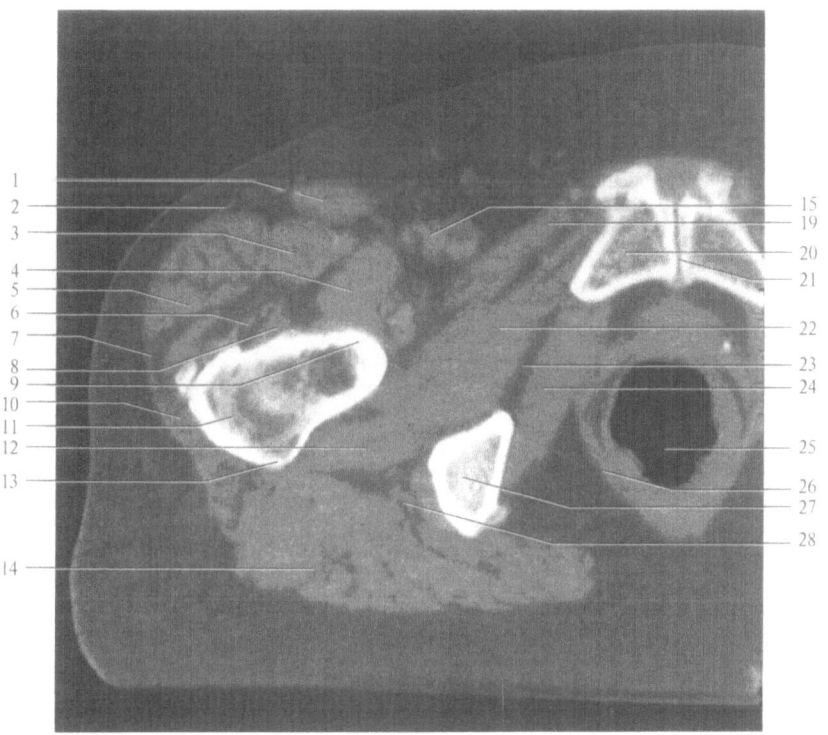

1 M. sartorius; *2* Fascia lata; *3* M. rectus femoris; *4* M. iliopsoas; *5* M. tensor fasciae latae; *6* M. vastus lateralis; *7* Tractus iliotibialis; *8* M. vastus intermedius; *9* Calcar femorale; *10* M. gluteus medius; *11* Trochanter major; *12* M. quadratus femoris; *13* Crista intertrochanterica; *14* M. gluteus maximus; *15* A. femoralis; *16* A. circumflexa femoris lateralis; *17* A. profunda femoris; *18* V. femoralis; *19* M. pectineus; *20* Os pubis; *21* Symphysis pubica; *22* M. obturator externus; *23* Foramen obturatum, Membrana obturatoria; *24* M. obturator internus; *25* Ampulla recti; *26* M. levator ani; *27* Tuber ischiadicum; *28* N. ischiadicus

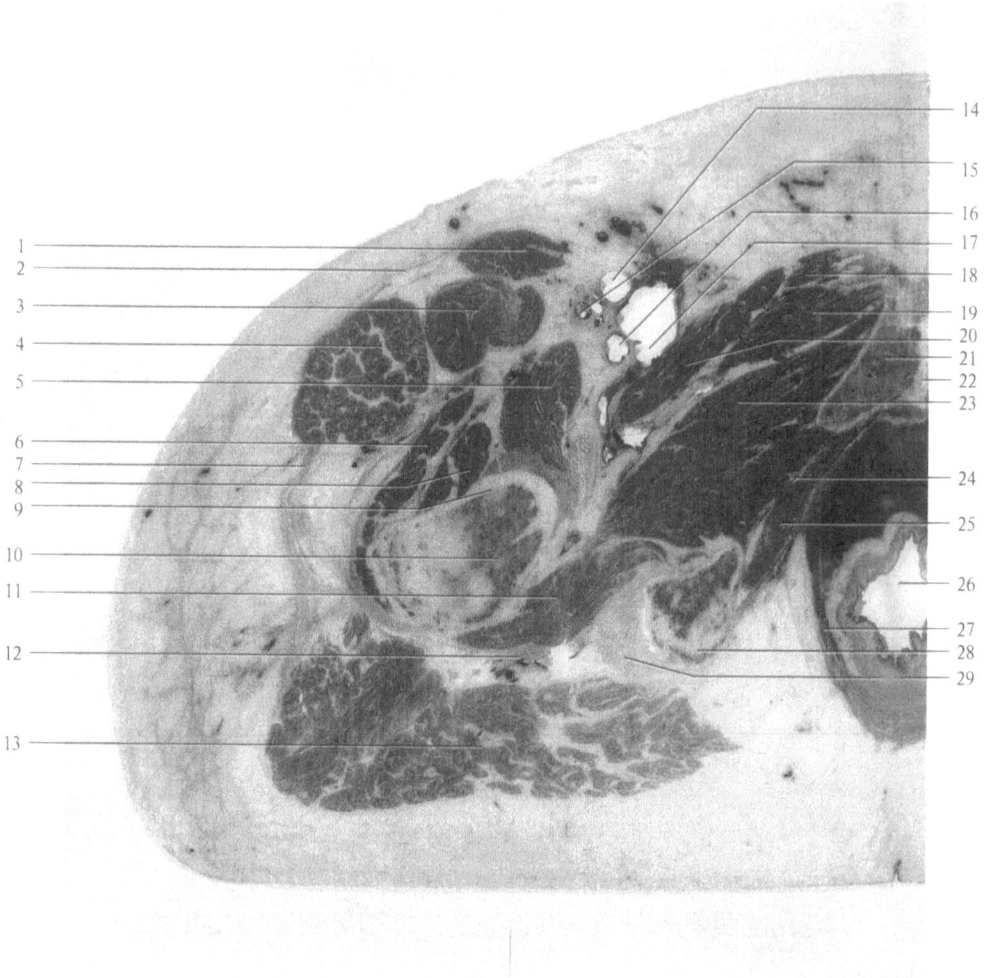

1
2
3
4
5
6
7
8
9
10
11
12
13

14
15
16
17
18
19
20
21
22
23
24
25
26
27
28
29

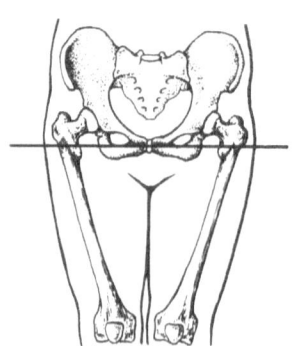

36

Plate 15

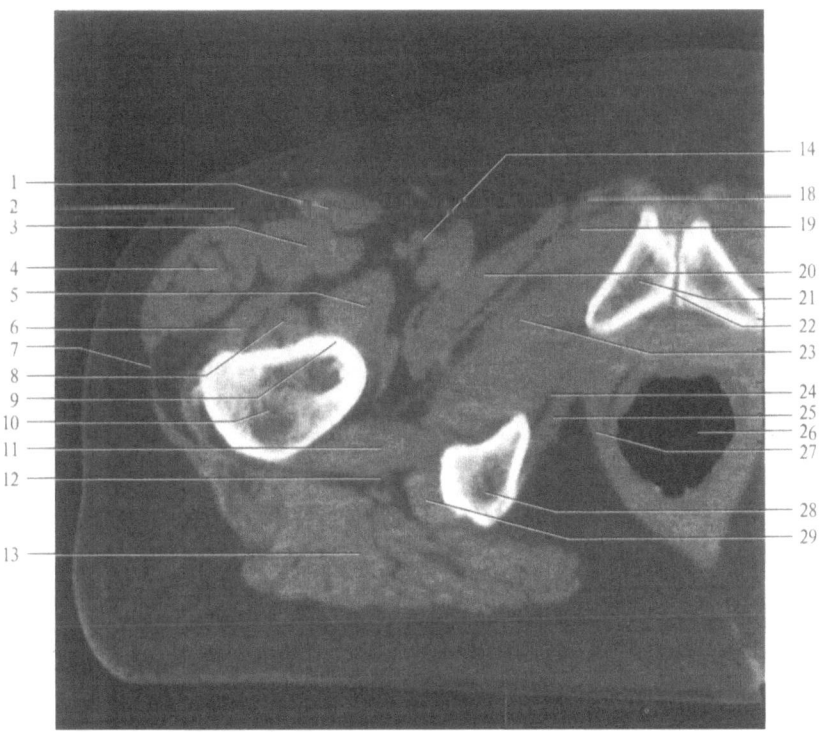

1 M. sartorius; 2 Fascia lata; 3 M. rectus femoris; 4 M. tensor fasciae latae; 5 M. iliopsoas; 6 M. vastus lateralis; 7 Tractus iliotibialis; 8 M. vastus intermedius; 9 Calcar femorale; 10 Femur; 11 M. quadratus femoris; 12 N. ischiadicus; 13 M. gluteus maximus; 14 A. femoralis; 15 A. circumflexa femoris lateralis; 16 A. profunda femoris; 17 V. femoralis; 18 M. adductor longus; 19 M. adductor brevis; 20 M. pectineus; 21 Os pubis; 22 Symphysis pubica; 23 M. obturator externus; 24 Foramen obturatum, Membrana obturatoria; 25 M. obturator internus; 26 Ampulla recti; 27 M. levator ani; 28 Tuber ischiadicum; 29 Origo mm. biceps femoris, semimembranosus et semitendinosus

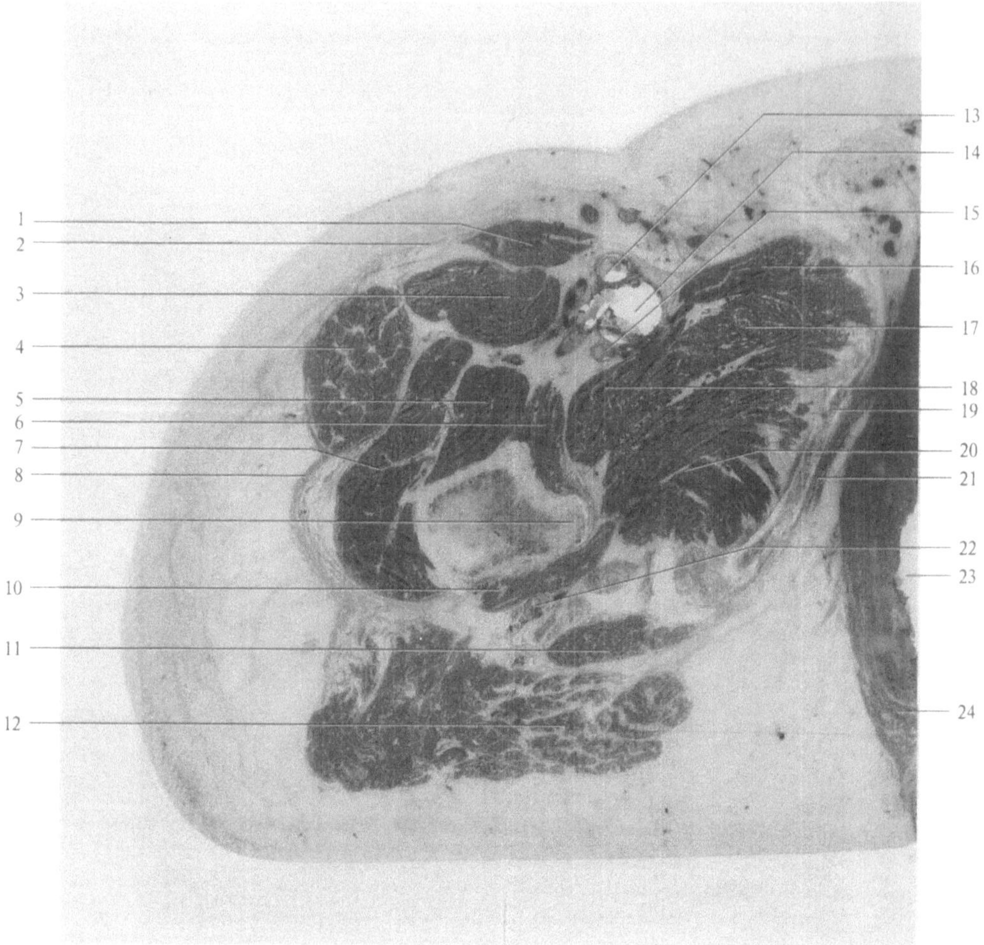

1
2
3
4
5
6
7
8
9
10
11
12

13
14
15
16
17
18
19
20
21
22
23
24

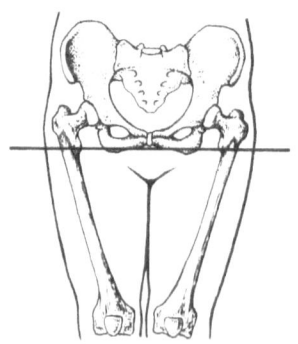

38

Plate 16

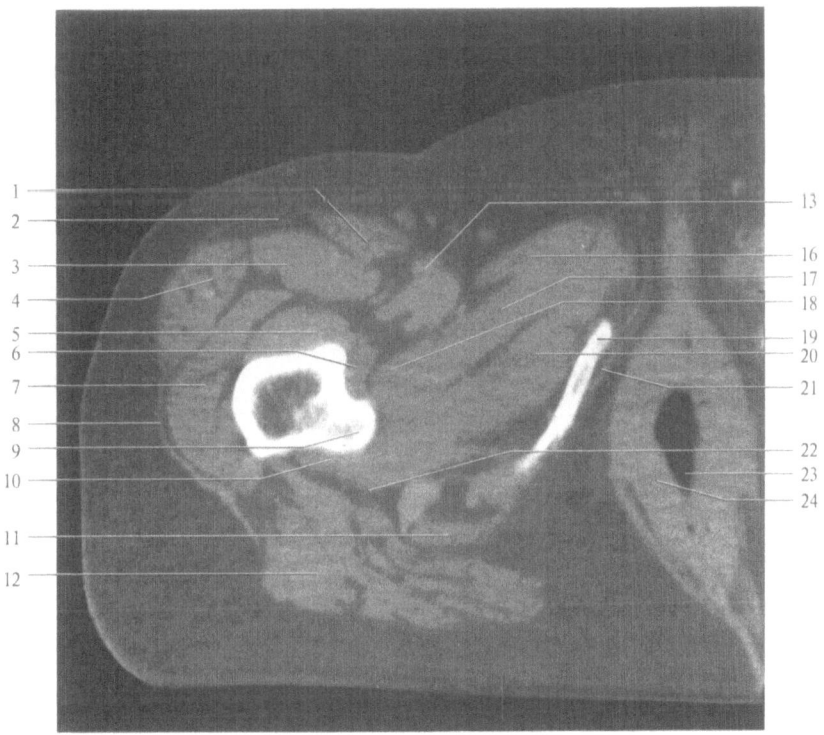

1 M. sartorius; *2* Fascia lata; *3* M. rectus femoris; *4* M. tensor fasciae latae; *5* M. vastus intermedius; *6* M. iliopsoas; *7* M. vastus lateralis; *8* Tractus iliotibialis; *9* Trochanter minor; *10* Insertio m. adductoris magni; *11* Caput commune mm. biceps femoris (caput longum) et semitendinosus; *12* M. gluteus maximus; *13* A. femoralis; *14* V. femoralis; *15* A. profunda femoris; *16* M. adductor longus; *17* M. adductor brevis; *18* M. pectineus; *19* Ramus inferior ossis pubis; *20* M. adductor magnus; *21* M. obturator internus; *22* N. ischiadicus; *23* Canalis analis; *24* M. sphincter ani externus

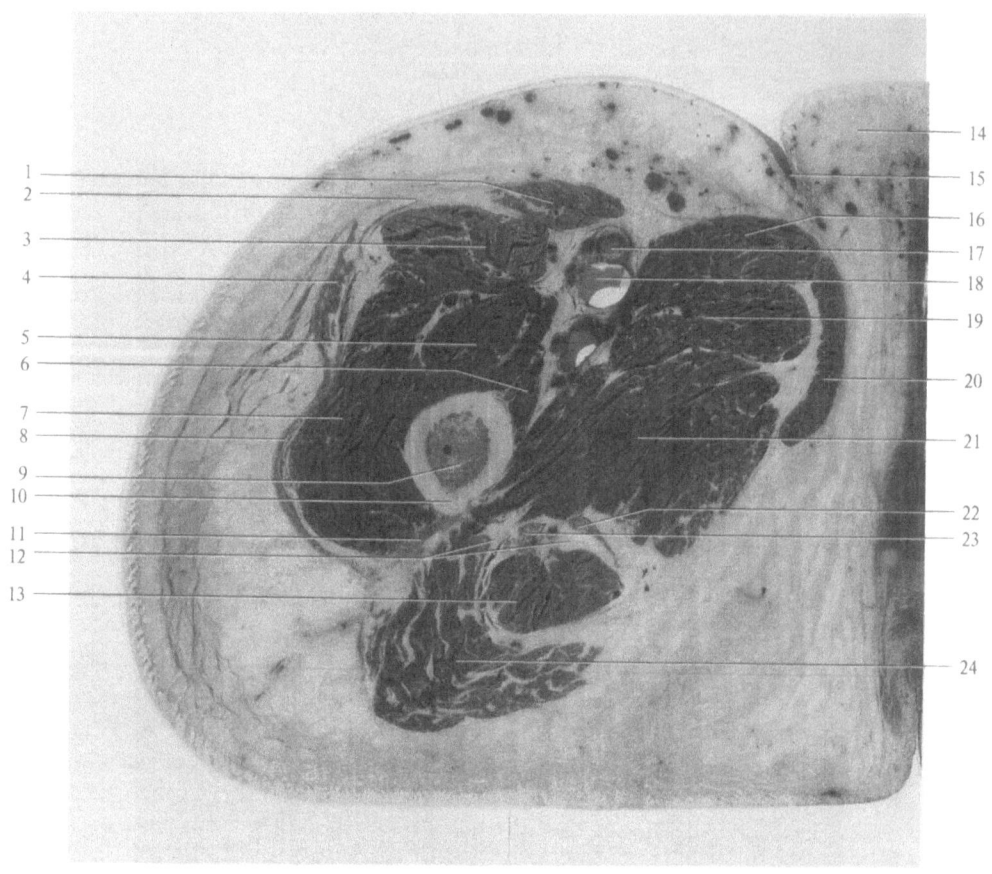

1
2
3
4
5
6
7
8
9
10
11
12
13

14
15
16
17
18
19
20
21
22
23

24

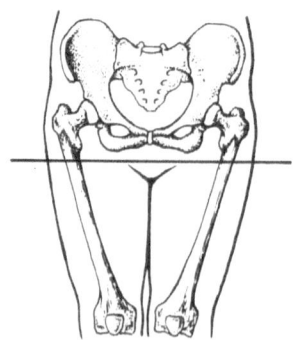

Plate 17

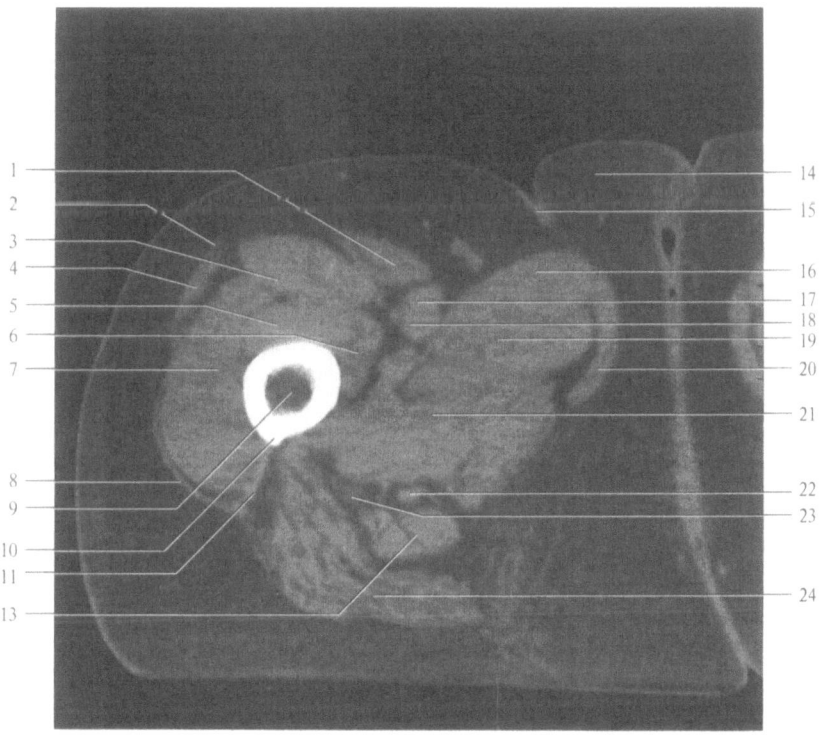

1 M. sartorius; *2* Fascia lata; *3* M. rectus femoris; *4* M. tensor fasciae latae; *5* M. vastus intermedius; *6* M. vastus medialis; *7* M. vastus lateralis; *8* Tractus iliotibialis; *9* Corpus ossis femoris; *10* Linea aspera; *11* Septum intermusculare femoris laterale; *12* Septum intermusculare femoris posterius; *13* Caput commune mm. biceps femoris (caput longum) et semitendinosus; *14* Labium majus pudendi; *15* Sulcus genitofemoralis; *16* M. adductor longus; *17* A. femoralis; *18* V. femoralis; *19* M. adductor brevis; *20* M. gracilis; *21* M. adductor magnus; *22* Tendo m. semimembranosi; *23* N. ischiadicus; *24* M. gluteus maximus

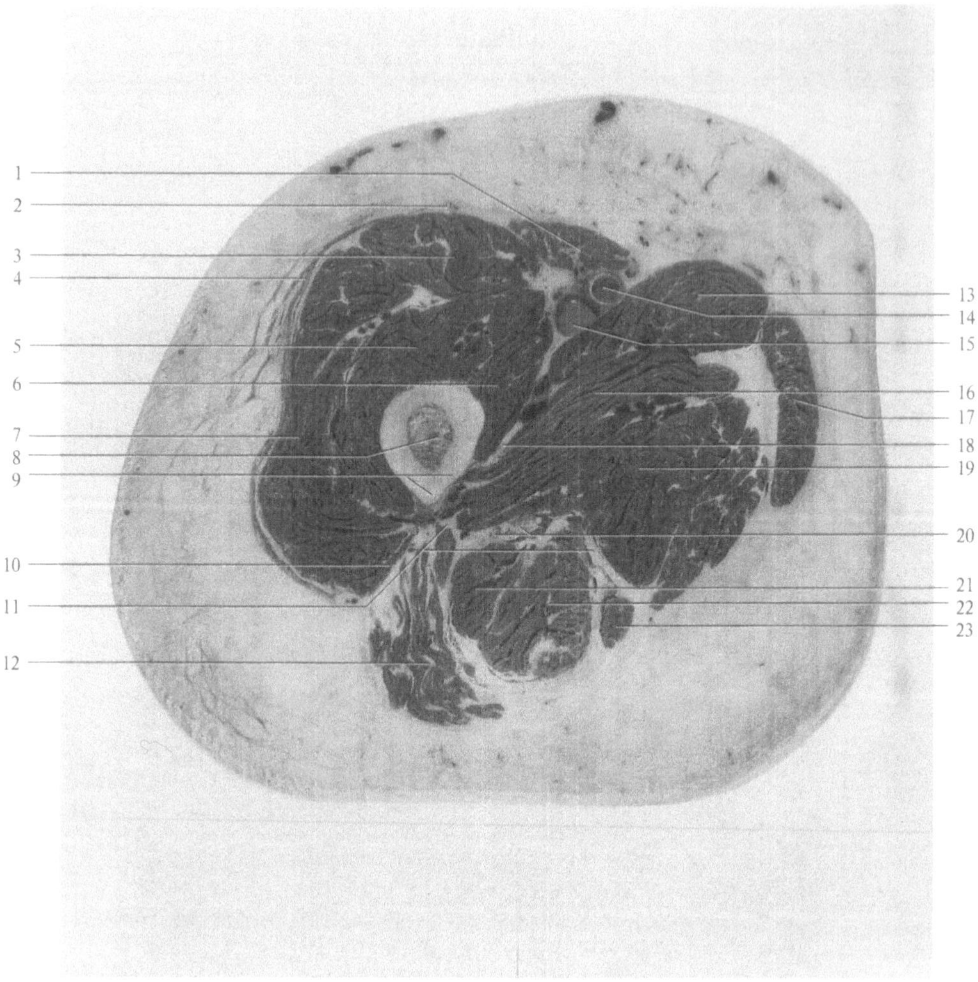

1
2
3
4
5
6
7
8
9
10
11
12

13
14
15
16
17
18
19

20
21
22
23

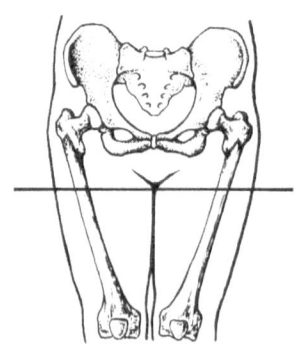

42

Plate 18

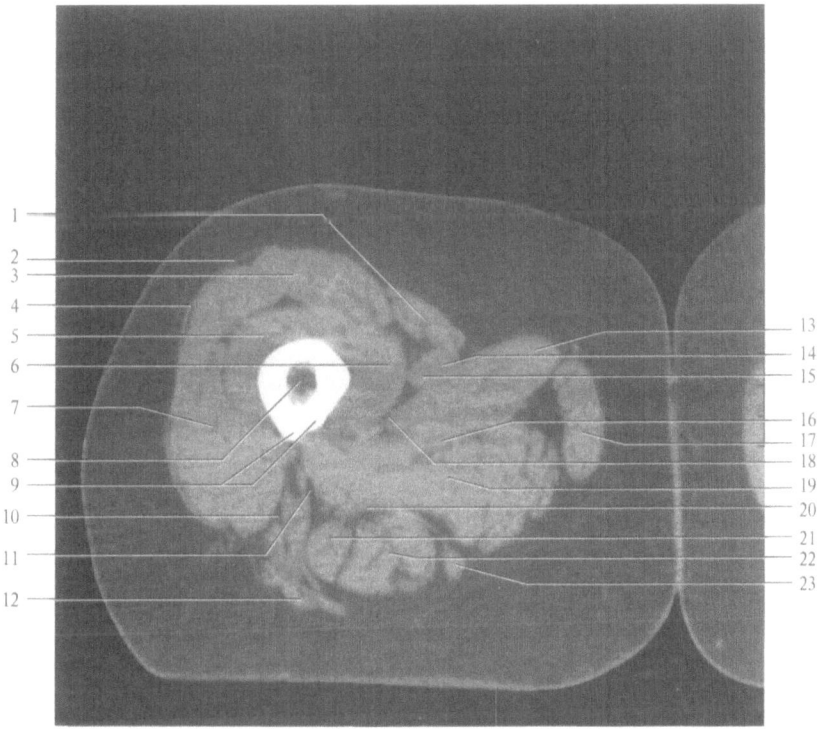

1 M. sartorius; 2 Fascia lata; 3 M. rectus femoris; 4 Tractus iliotibialis; 5 M. vastus intermedius; 6 M. vastus medialis; 7 M. vastus lateralis; 8 Corpus ossis femoris; 9 Linea aspera; 10 Septum intermusculare femoris laterale; 11 Septum intermusculare femoris posterius; 12 M. gluteus maximus; 13 M. adductor longus; 14 A. femoralis; 15 V. femoralis; 16 M. adductor brevis; 17 M. gracilis; 18 Septum intermusculare femoris mediale; 19 M. adductor magnus; 20 N. ischiadicus; 21 Caput longum m. bicipitis femoris; 22 M. semitendinosus; 23 M. semimembranosus

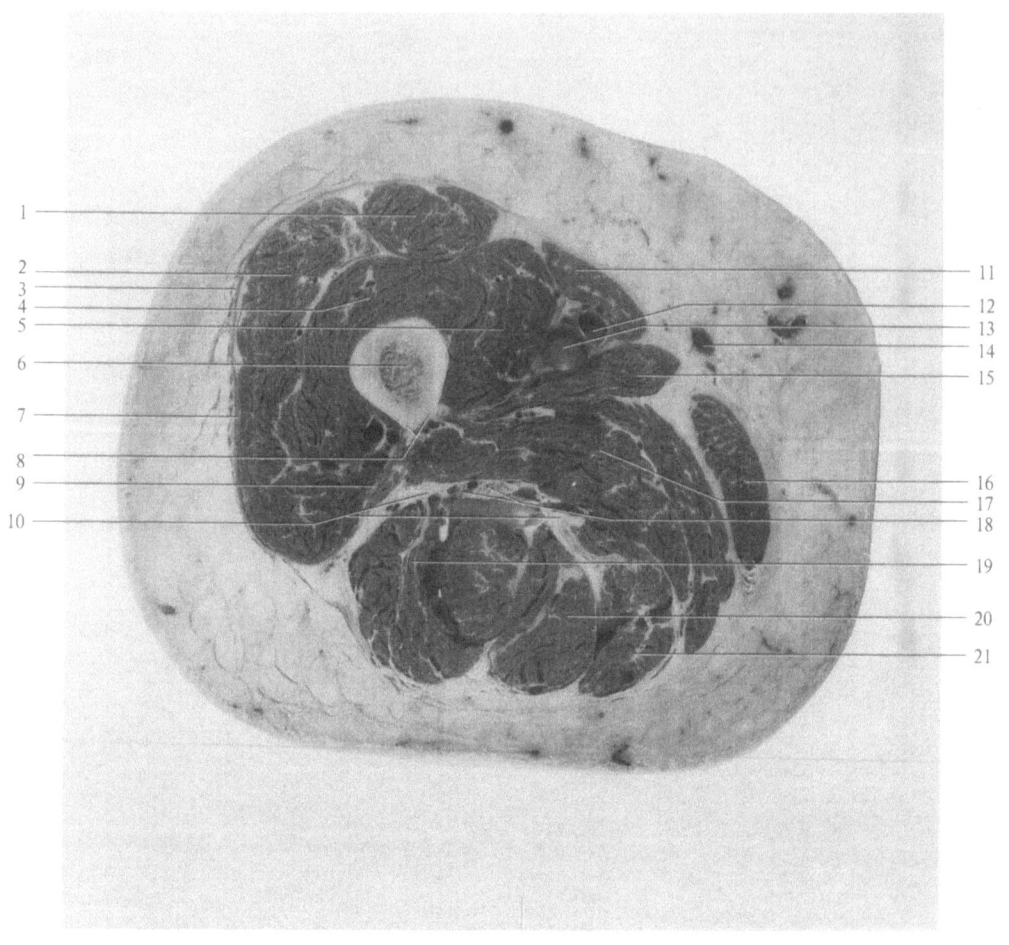

1 —————————

2 —————————
3 —————————
4 —————————
5 —————————

6 —————————

7 —————————

8 —————————
9 —————————

10 —————————

————————— 11

————————— 12
————————— 13
————————— 14
————————— 15

————————— 16
————————— 17
————————— 18

————————— 19

————————— 20
————————— 21

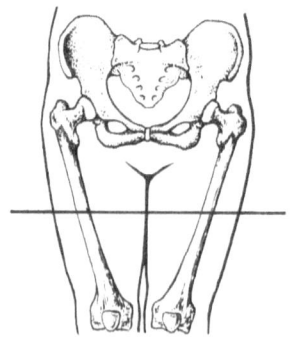

Plate 19

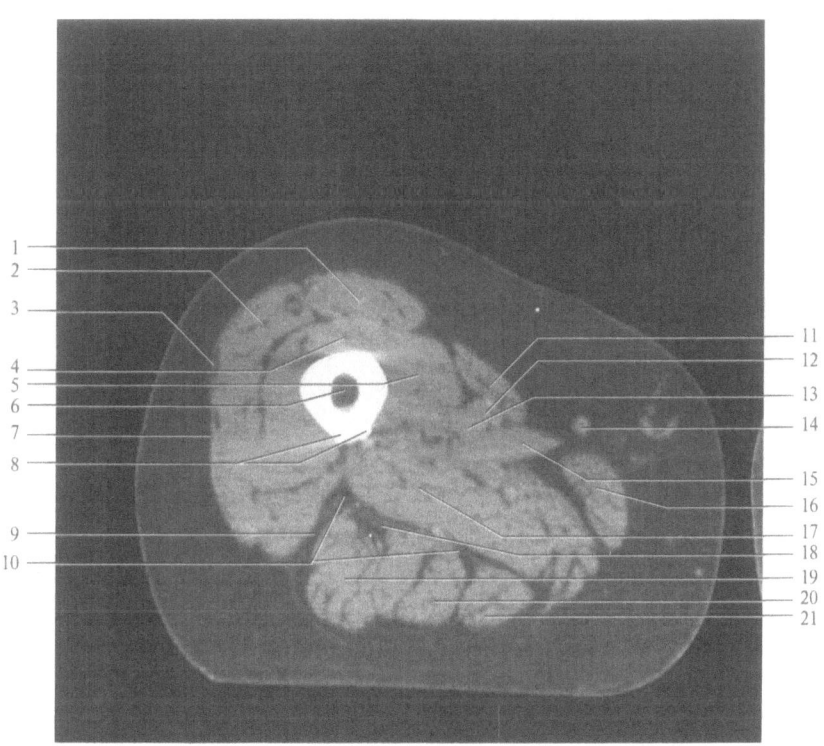

1 M. rectus femoris; 2 M. vastus lateralis; 3 Tractus iliotibialis; 4 M. vastus intermedius; 5 M. vastus medialis; 6 Corpus ossis femoris; 7 Fascia lata; 8 Linea aspera; 9 Septum intermusculare femoris laterale; 10 Septum intermusculare femoris posterius; 11 M. sartorius; 12 A. femoralis; 13 V. femoralis; 14 V. saphena magna; 15 M. adductor longus; 16 M. gracilis; 17 M. adductor magnus; 18 N. ischiadicus; 19 Caput longum m. bicipitis femoris; 20 M. semitendinosus; 21 M. semimembranosus

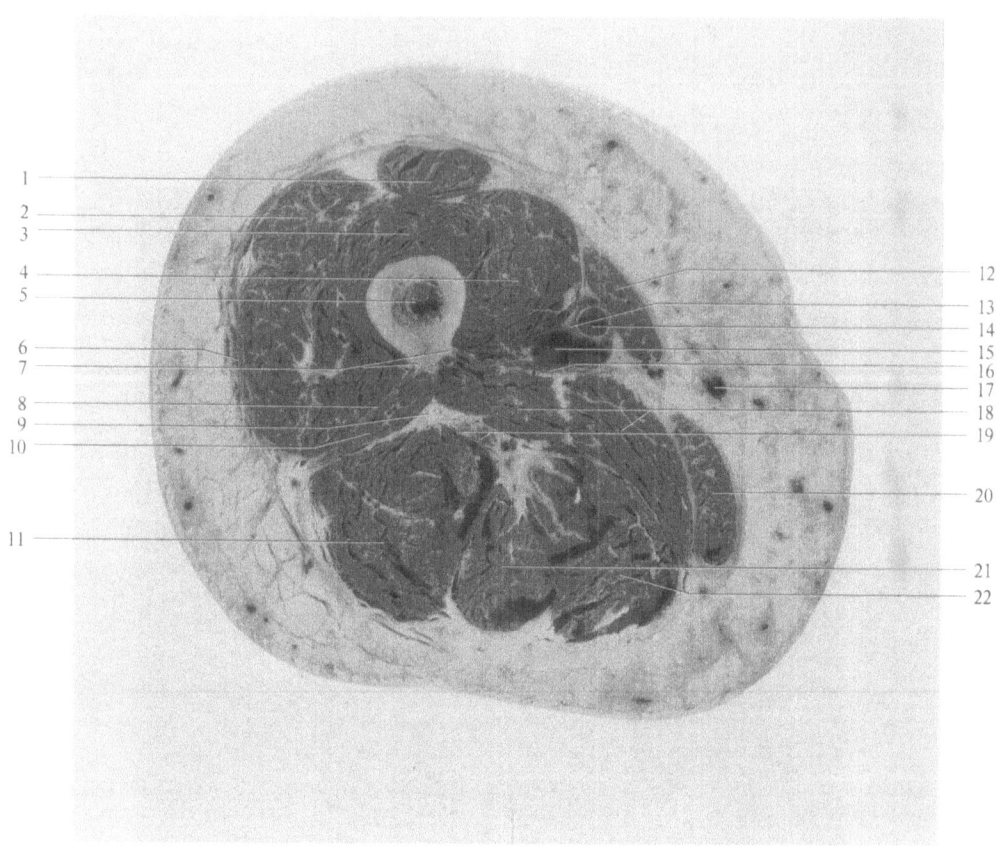

1 —————
2 —————
3 —————

4 —————
5 —————

6 —————
7 —————

8 —————
9 —————
10 —————

11 —————

12 —————
13 —————
14 —————
15 —————
16 —————
17 —————
18 —————
19 —————

20 —————

21 —————
22 —————

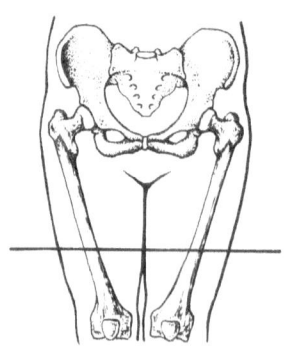

Plate 20

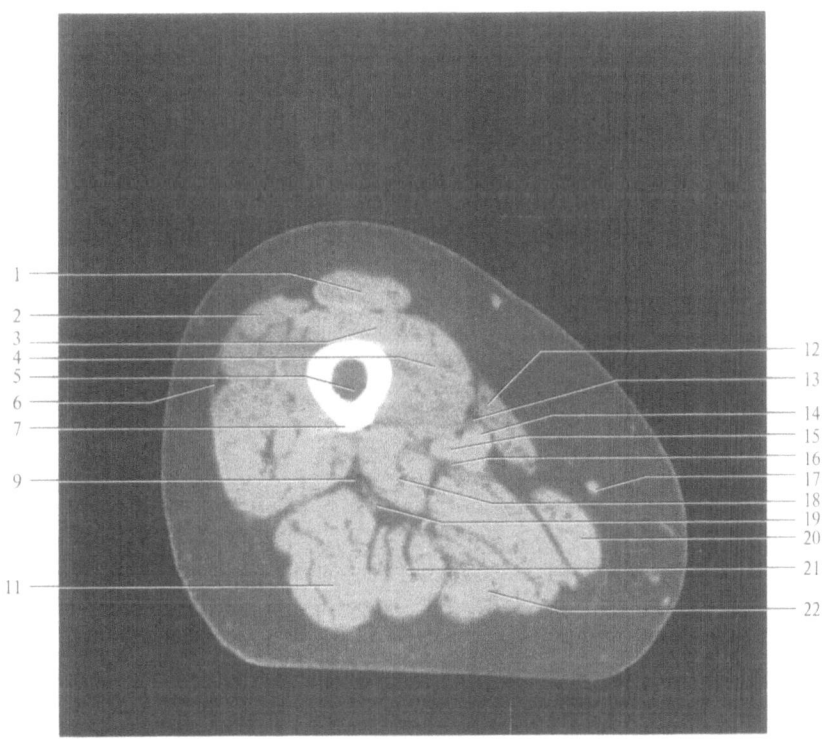

1 M. rectus femoris; 2 M. vastus lateralis; 3 M. vastus intermedius; 4 M. vastus medialis; 5 Corpus ossis femoris; 6 Fascia lata; 7 Linea aspera; 8 Septum intermusculare
femoris laterale; 9 Septum intermusculare femoris posterius; 10 Caput breve m. bicipitis femoris; 11 Caput longum m. bicipitis femoris; 12 M. sartorius; 13 Canalis adducto-
rius; 14 A. femoralis; 15 V. femoralis; 16 Membrana vasto-adductoria; 17 V. saphena
magna; 18 M. adductor magnus; 19 N. ischiadicus; 20 M. gracilis; 21 M. semitendinosus;
22 M. semimembranosus

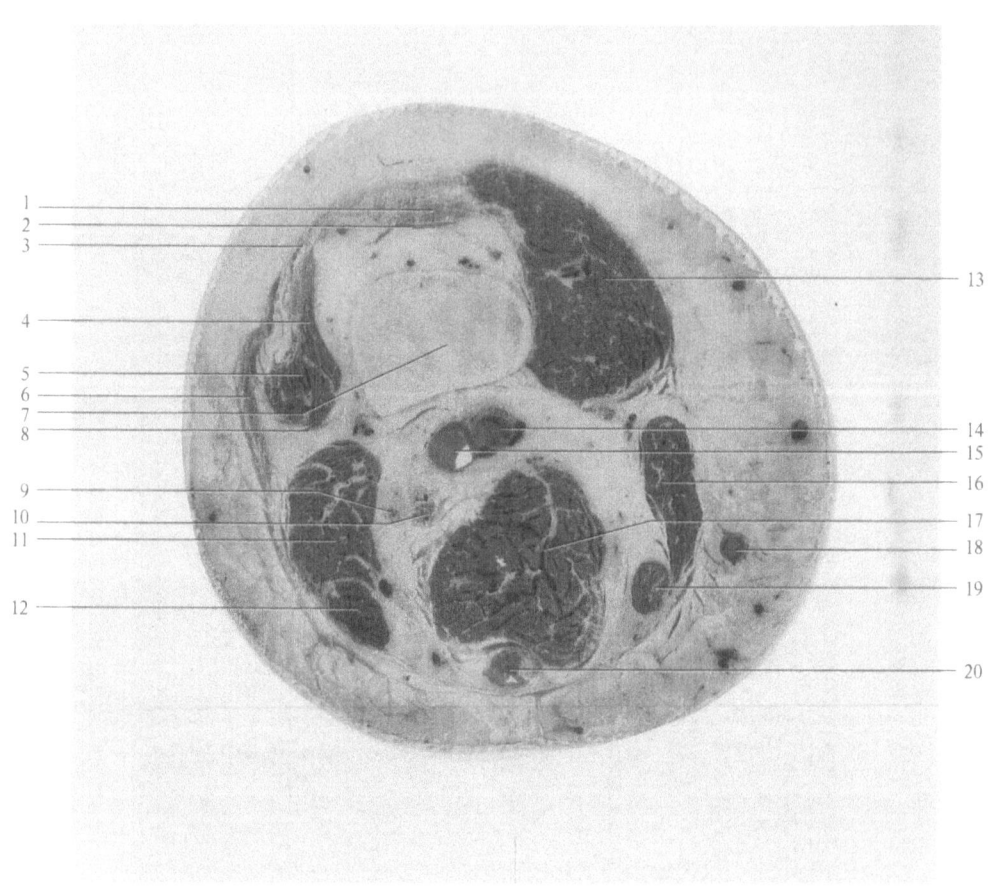

1
2
3

4

5
6
7
8

9
10
11

12

13

14
15
16
17
18
19

20

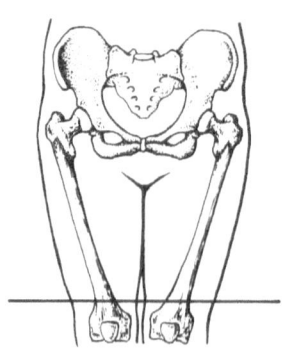

48

Plate 21

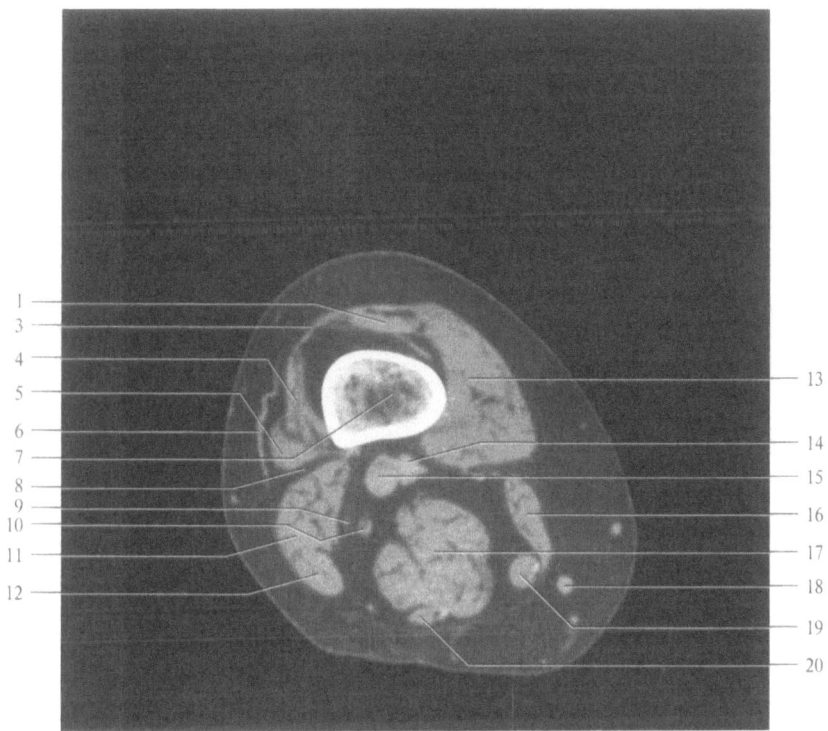

1 Tendo m. quadricipitis femoris; 2 Bursa suprapatellaris; 3 Retinaculum patellae laterale; 4 M. vastus intermedius; 5 M. vastus lateralis; 6 Tractus iliotibialis; 7 Metaphysis
ossis femoris; 8 Septum intermusculare femoris laterale; 9 N. peroneus communis;
10 N. tibialis; 11 Caput breve m. bicipitis femoris; 12 Caput longum m. bicipitis femoris; 13 M. vastus medialis; 14 A. poplitea; 15 V. poplitea; 16 M. sartorius; 17 M. semimembranosus; 18 V. saphena magna; 19 M. gracilis; 20 M. semitendinosus

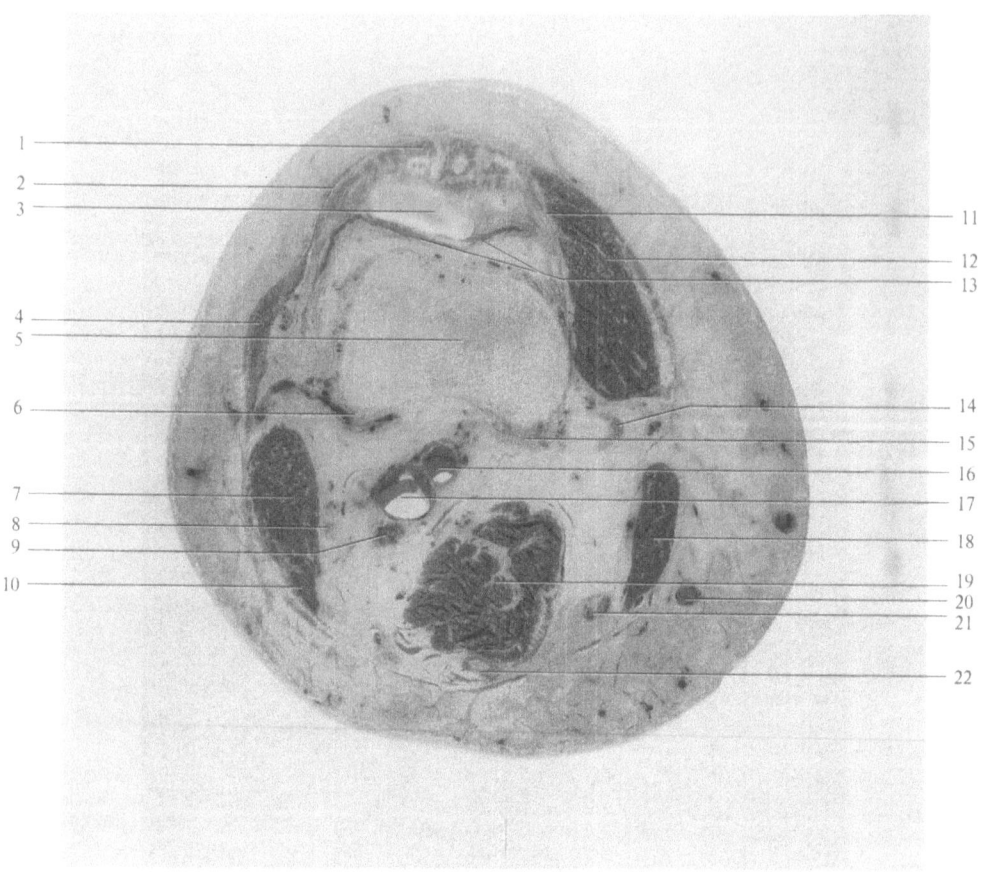

1
2
3

4
5

6

7
8
9
10

11

12
13

14
15
16
17
18

19
20
21

22

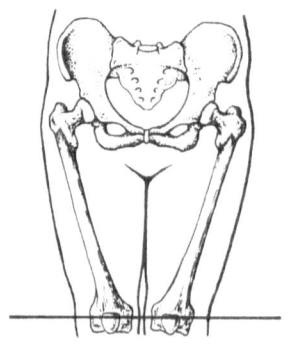

50

Plate 22

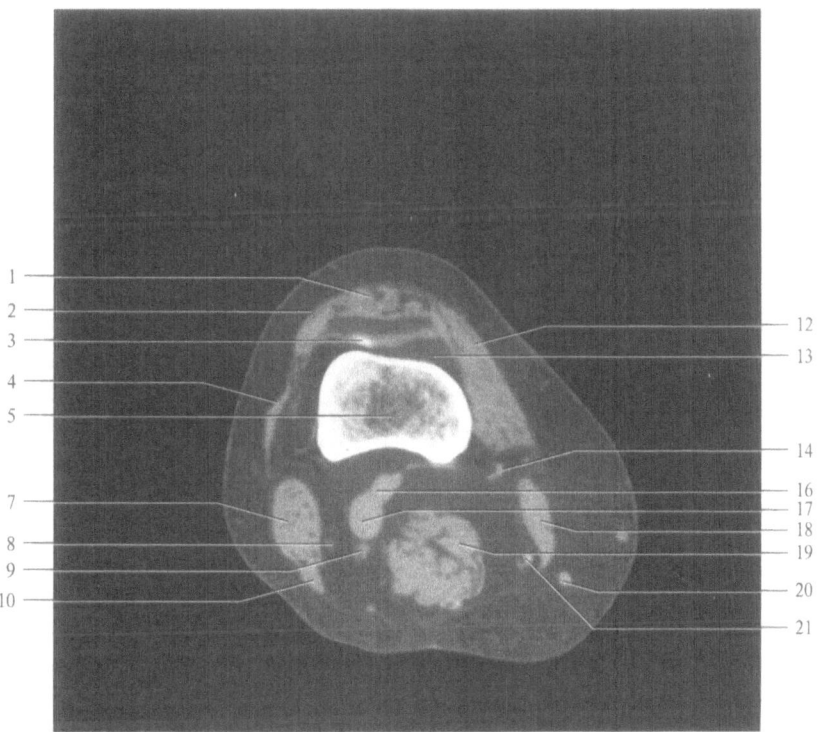

1 Tendo m. quadricipitis femoris; 2 Retinaculum patellae laterale; 3 Patella; 4 Tractus iliotibialis; 5 Metaphysis ossis femoris; 6 Origo m. plantaris; 7 Caput breve m. bicipitis femoris; 8 N. peroneus communis; 9 N. tibialis; 10 Caput longum m. bicipitis femoris; 11 Retinaculum patellae mediale; 12 M. vastus medialis; 13 Cavitas articularis; 14 Tendo m. adductoris magni; 15 Origo capitis medialis m. gastrocnemii; 16 A. poplitea; 17 V. poplitea; 18 M. sartorius; 19 M. semimembranosus; 20 V. saphena magna; 21 Tendo m. gracilis; 22 M. semitendinosus

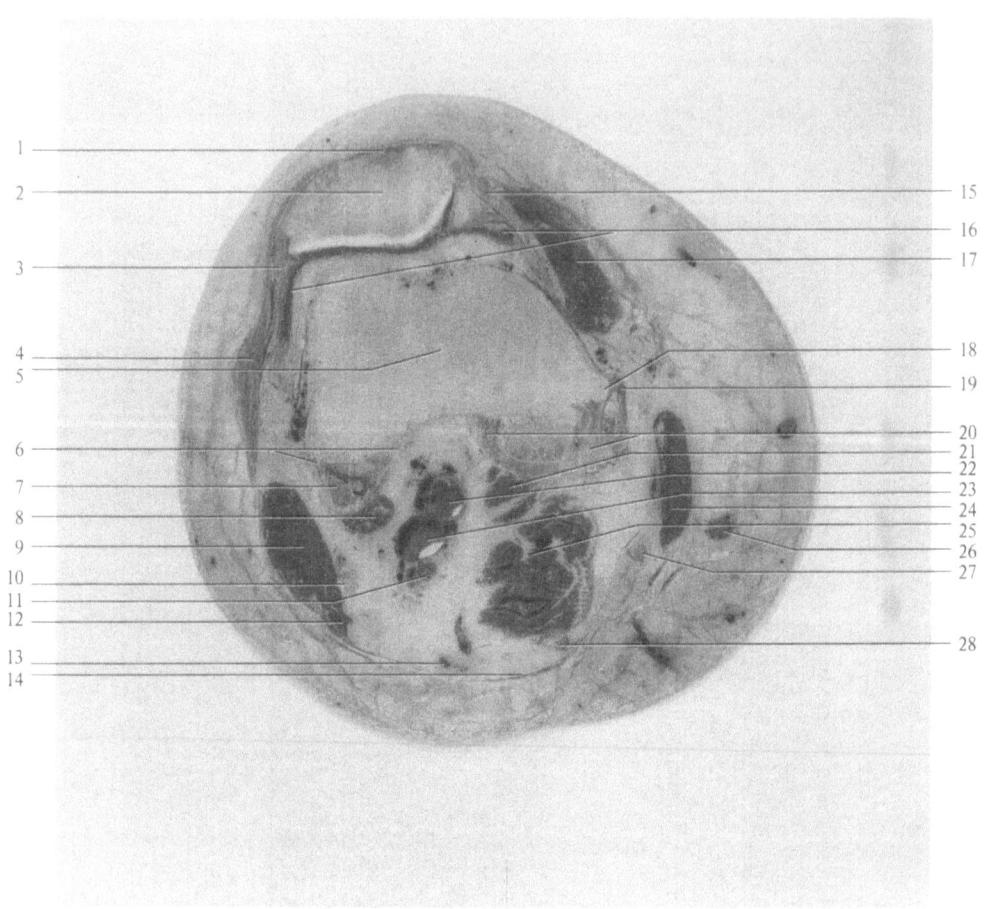

1
2
3
4
5
6
7
8
9
10
11
12
13
14

15
16
17
18
19
20
21
22
23
24
25
26
27
28

52

Plate 23

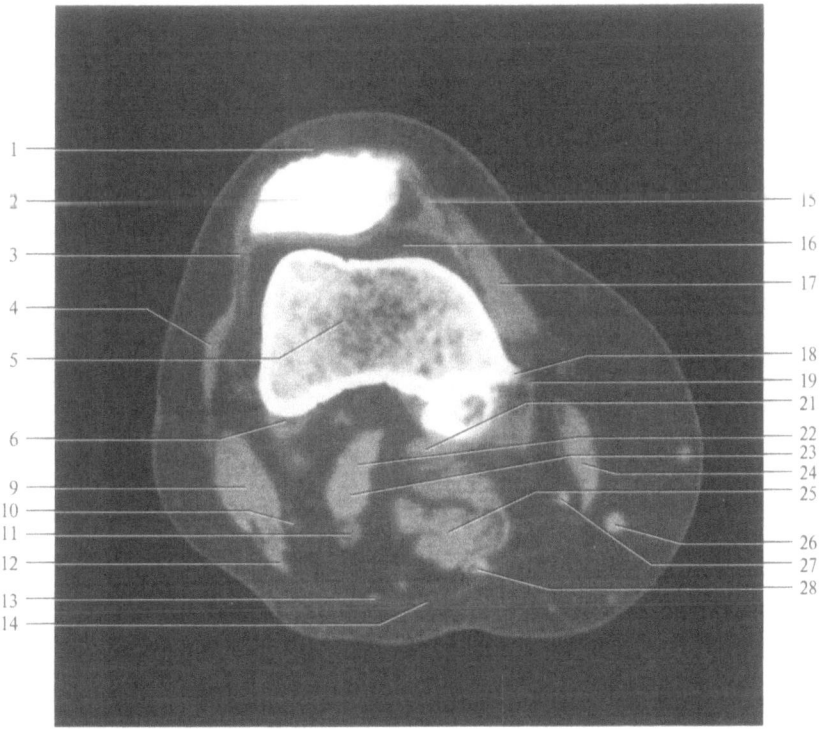

1 Tendo m. quadricipitis femoris; *2* Patella; *3* Retinaculum patellae laterale; *4* Tractus iliotibialis; *5* Epiphysis distalis ossis femoris; *6* Origo capitis lateralis m. gastrocnemii; *7* Fabella; *8* M. plantaris; *9* Caput breve m. bicipitis femoris; *10* N. peroneus communis; *11* N. tibialis; *12* Caput longum m. bicipitis femoris; *13* V. saphena parva; *14* Fascia poplitea; *15* Retinaculum patellae mediale; *16* Cavitas articularis; *17* M. vastus medialis; *18* Tuberculum adductorium; *19* Insertio m. adductoris magni; *20* Origo capitis medialis m. gastrocnemii; *21* Caput mediale m. gastrocnemii; *22* A. poplitea; *23* V. poplitea; *24* M. sartorius; *25* M. semimembranosus; *26* V. saphena magna; *27* Tendo m. gracilis; *28* M. semitendinosus

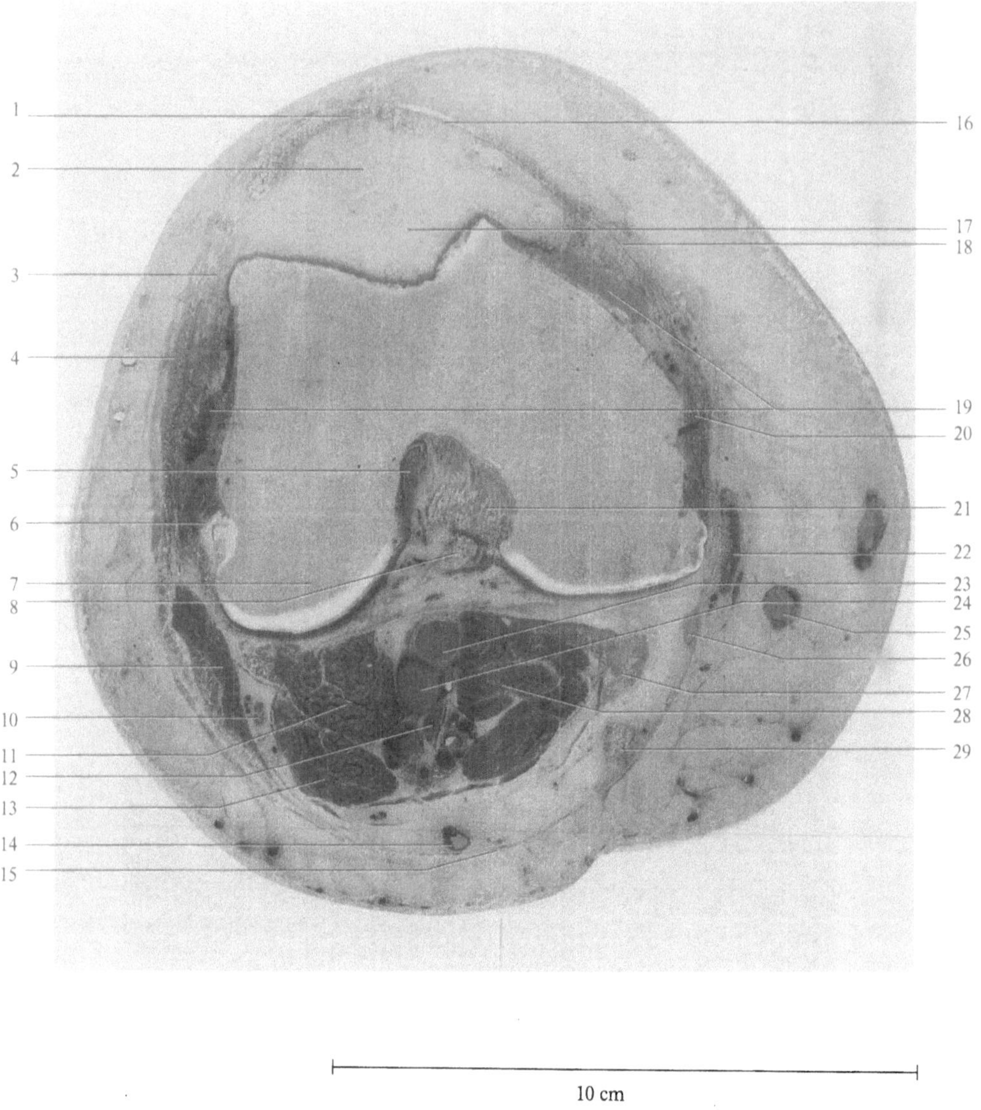

1

2

3

4

5

6

7

8

9

10

11

12

13

14

15

16

17

18

19

20

21

22

23

24

25

26

27

28

29

10 cm

54

Plate 24

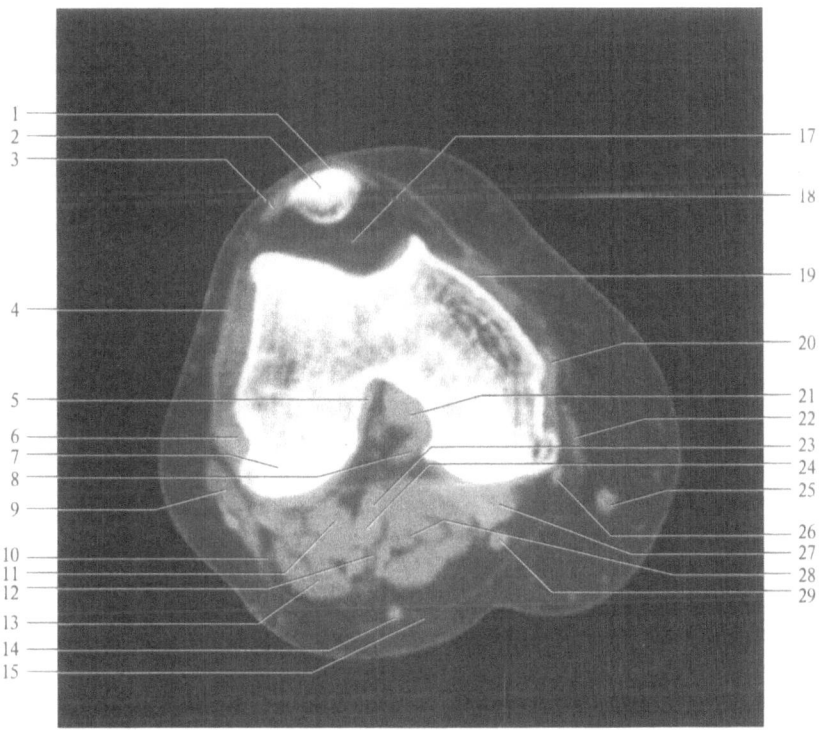

1 Tendo m. quadricipitis femoris; *2* Patella; *3* Retinaculum patellae laterale; *4* Tractus iliotibialis; *5* Lig. cruciatum anterius; *6* Lig. collaterale fibulare; *7* Condylus lateralis ossis femoris; *8* Lig. meniscofemorale posterius; *9* M. biceps femoris; *10* N. peroneus communis; *11* M. plantaris; *12* N. tibialis; *13* Caput laterale m. gastrocnemii; *14* V. saphena parva; *15* Fascia poplitea; *16* Bursa subcutanea prepatellaris; *17* Corpus adiposum infrapatellare; *18* Retinaculum patellae mediale; *19* Cavitas articularis; *20* Lig. collaterale tibiale; *21* Lig. cruciatum posterius; *22* M. sartorius; *23* A. poplitea; *24* V. poplitea; *25* V. saphena magna; *26* Tendo m. gracilis; *27* Tendo m. semimembranosi; *28* Caput mediale m. gastrocnemii; *29* Tendo m. semitendinosi

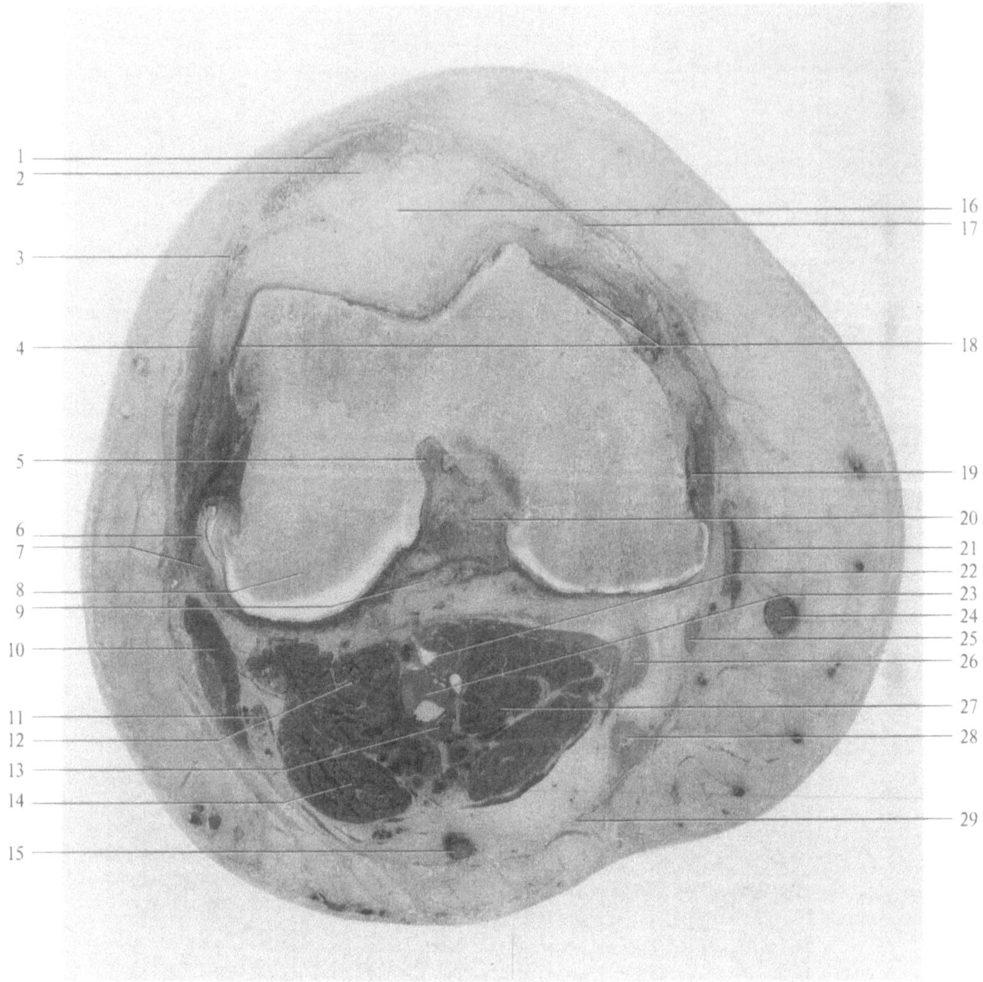

1

2

3

4

5

6

7

8

9

10

11

12

13

14

15

16

17

18

19

20

21

22

23

24

25

26

27

28

29

56

Plate 25

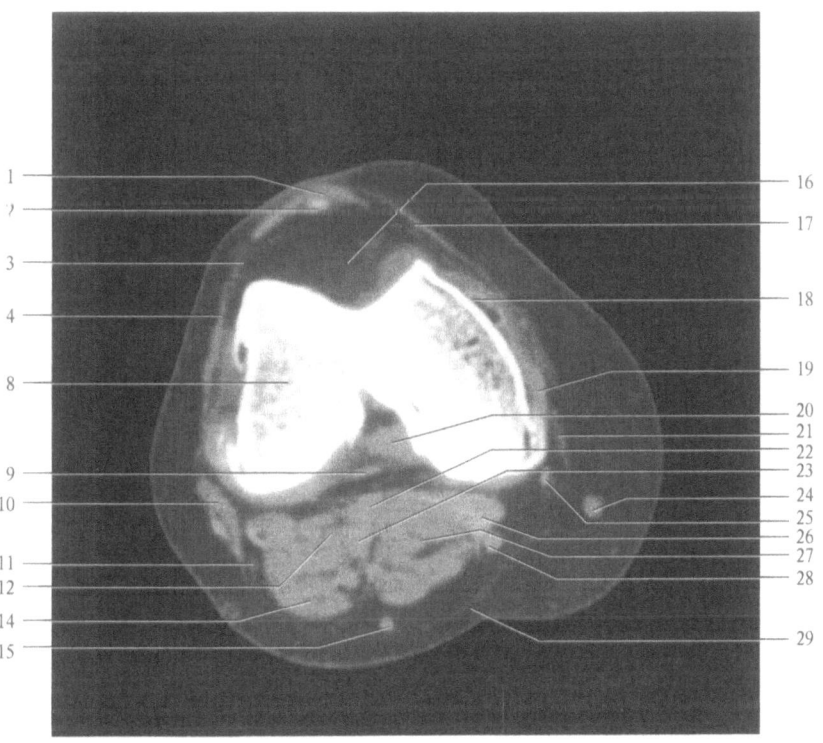

1 Tendo m. quadricipitis femoris; *2* Apex patellae; *3* Retinaculum patellae laterale; *4* Tractus iliotibialis; *5* Lig. cruciatum anterius; *6* Tendo m. poplitei; *7* Lig. collaterale fibulare; *8* Condylus lateralis ossis femoris; *9* Lig. meniscofemorale posterius; *10* M. biceps femoris; *11* N. peroneus communis; *12* M. plantaris; *13* N. tibialis; *14* Caput laterale m. gastrocnemii; *15* V. saphena parva; *16* Corpus adiposum infrapatellare; *17* Retinaculum patellae mediale; *18* Cavitas articularis; *19* Lig. collaterale tibiale; *20* Lig. cruciatum posterius; *21* M. sartorius; *22* A. poplitea; *23* V. poplitea; *24* V. saphena magna; *25* Tendo m. gracilis; *26* Tendo m. semimembranosi; *27* Caput mediale m. gastrocnemii; *28* Tendo m. semitendinosi; *29* Fascia poplitea

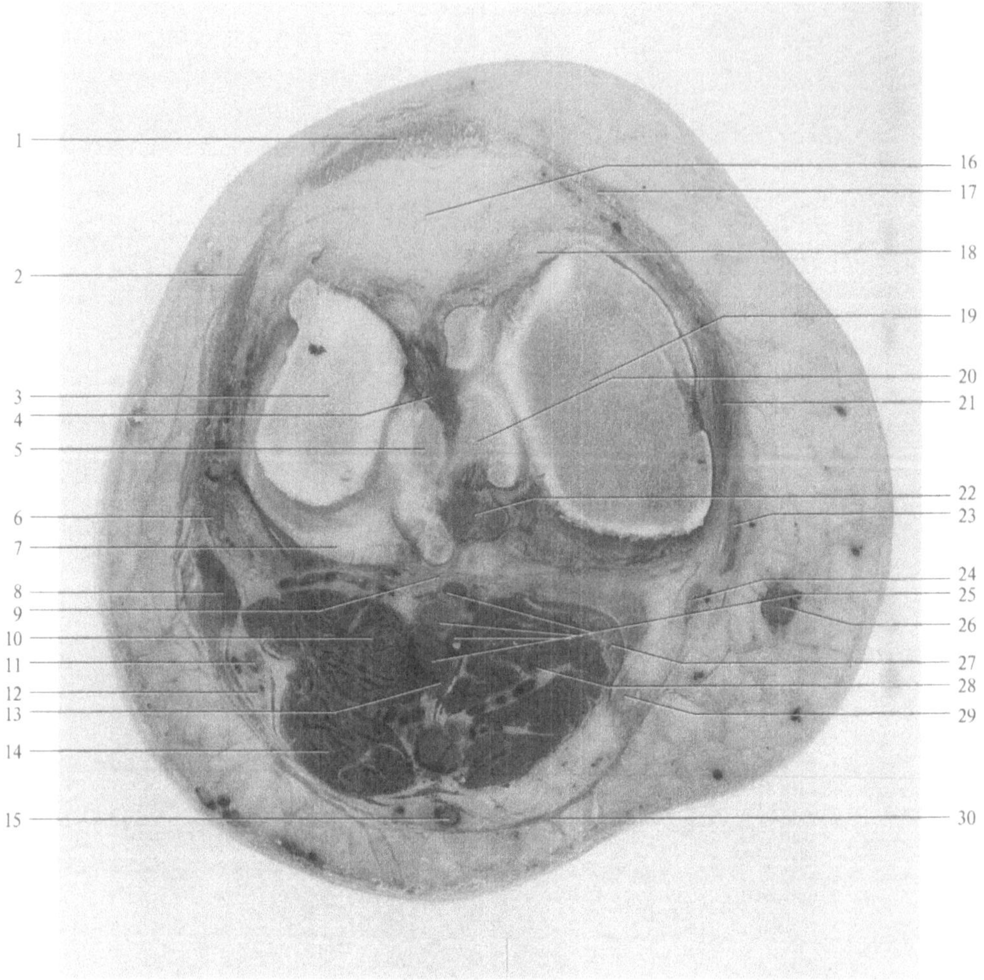

1

2

3
4
5

6
7

8
9
10
11
12
13

14

15

16
17

18

19

20
21

22
23

24
25
26
27
28
29

30

58

Plate 26

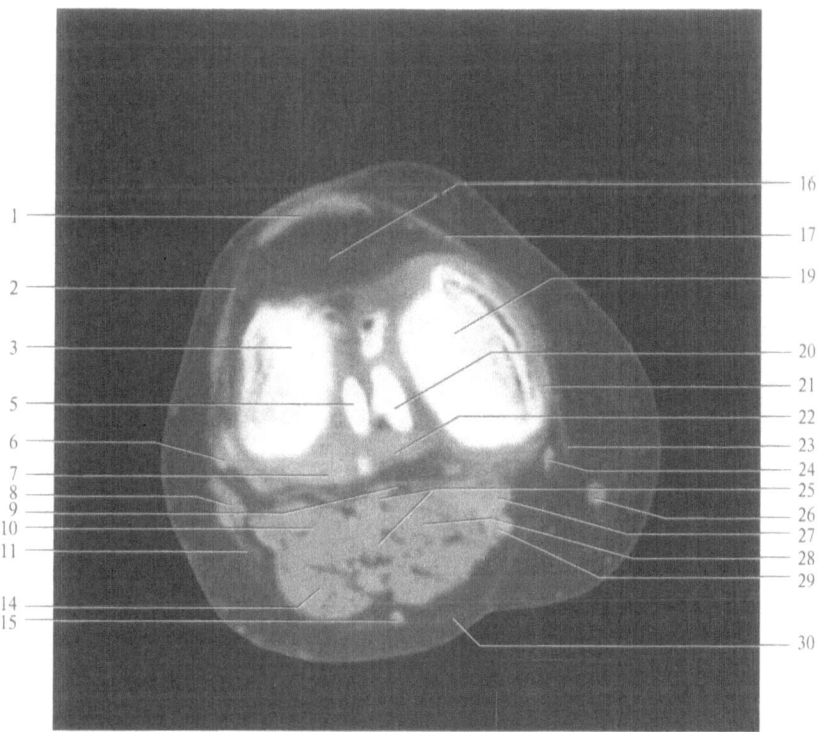

1 Lig. patellae; *2* Retinaculum patellae laterale; *3* Condylus lateralis ossis femoris; *4* Lig. cruciatum anterius; *5* Tuberculum intercondylare laterale tibiae; *6* Lig. collaterale fibulare; *7* Meniscus lateralis; *8* M. biceps femoris; *9* Capsula articularis; *10* M. plantaris; *11* N. peroneus profundus; *12* N. peroneus superficialis; *13* N. tibialis; *14* Caput laterale m. gastrocnemii; *15* V. saphena parva; *16* Corpus adiposum infrapatellare; *17* Retinaculum patellae mediale; *18* Meniscus medialis; *19* Condylus medialis ossis femoris; *20* Tuberculum intercondylare mediale tibiae; *21* Lig. collaterale tibiale; *22* Lig. cruciatum posterius; *23* Tendo m. sartorii; *24* Tendo m. gracilis; *25* Vasa tibialia; *26* V. saphena magna; *27* Tendo m. semimembranosi; *28* Caput mediale m. gastrocnemii; *29* Tendo m. semitendinosi; *30* Fascia poplitea

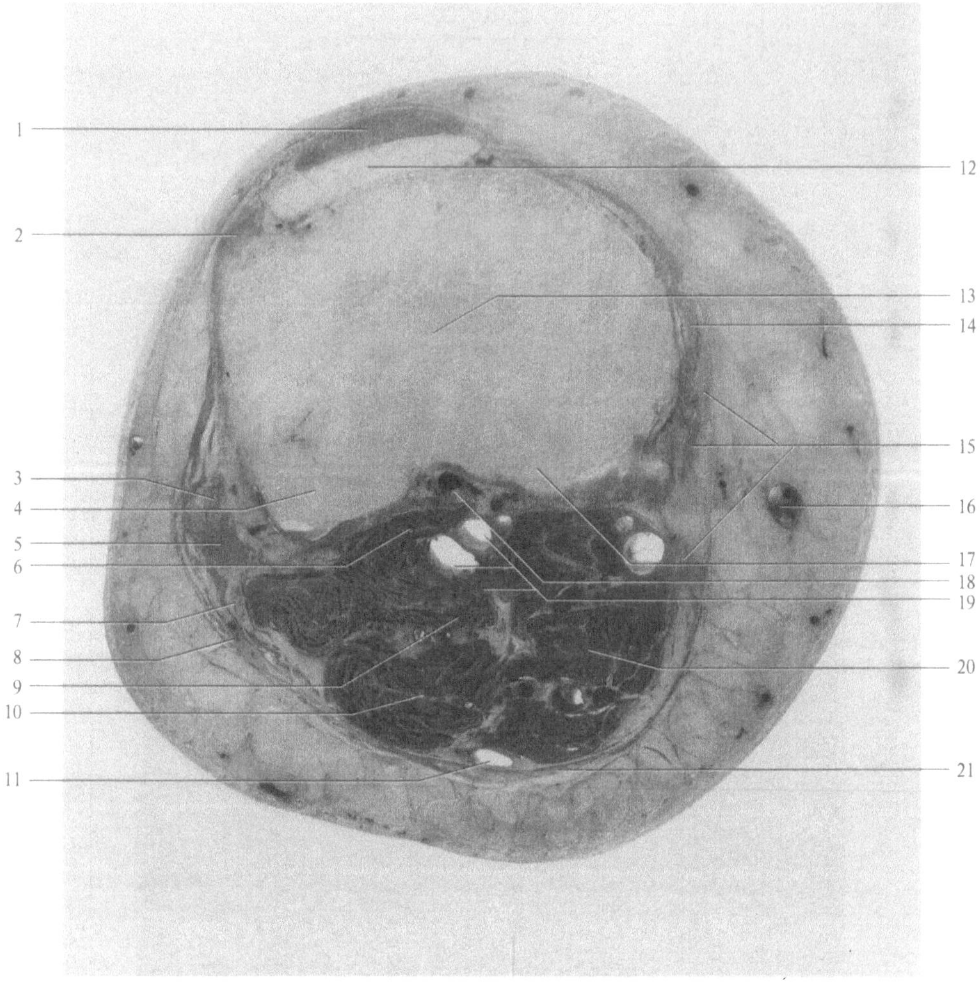

1

2

3
4
5
6

7
8
9
10

11

12

13
14

15

16

17
18
19

20

21

60

Plate 27

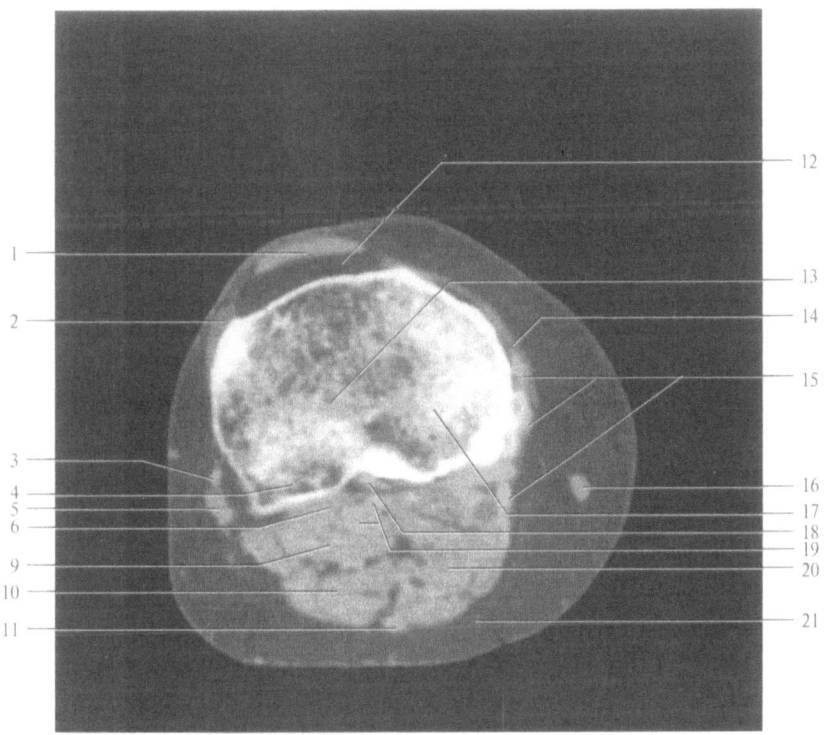

1 Lig. patellae; *2* Tendines mm. tibialis anterior et extensor digitorum longus; *3* Lig. collaterale fibulare; *4* Condylus lateralis tibiae; *5* Tendo m. bicipitis femoris; *6* M. popliteus; *7* N. peroneus profundus; *8* N. peroneus superficialis; *9* M. plantaris; *10* Caput laterale m. gastrocnemii; *11* V. saphena parva; *12* Corpus adiposum infrapatellare; *13* Tibia; *14* Lig. collaterale tibiale; *15* Pes anserinus superficialis (Tendines mm. sartorius, gracilis, semitendinosus); *16* V. saphena magna; *17* Condylus medialis tibiae; *18* A. tibialis anterior; *19* N. tibialis et Vasa tibialia posteriora; *20* Caput mediale m. gastrocnemii; *21* Fascia poplitea

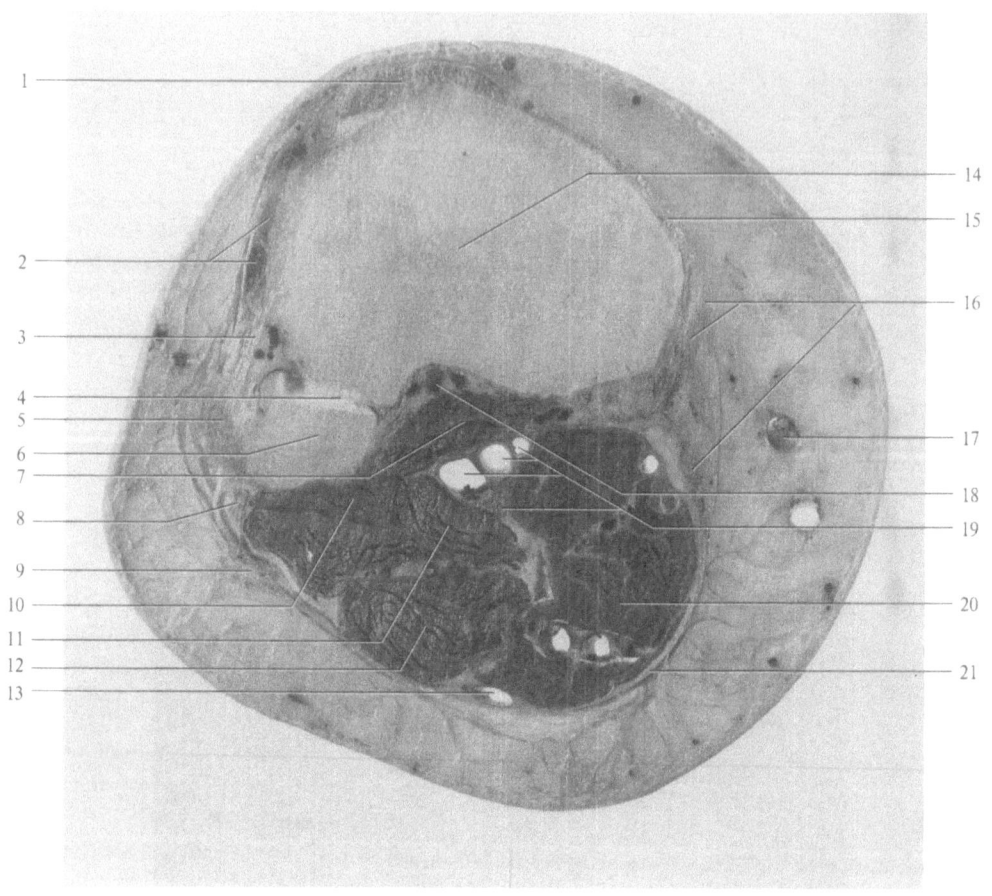

1
2
3
4
5
6
7
8
9
10
11
12
13

14
15
16
17
18
19
20
21

62

Plate 28

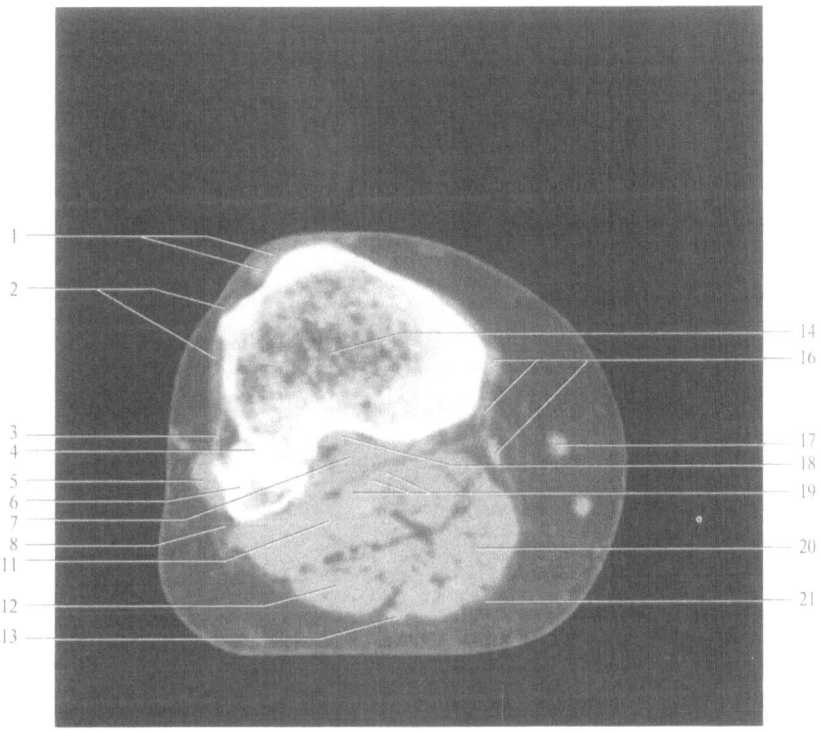

1 Lig. patellae; *2* Tendines mm. tibialis anterior et extensor digitorum longus; *3* Lig. tibiofibulare; *4* Articulatio tibiofibularis; *5* Origo m. peronei longi; *6* Caput fibulae;
7 M. popliteus; *8* N. peroneus profundus; *9* N. peroneus superficialis; *10* M. soleus;
11 M. plantaris; *12* Caput laterale m. gastrocnemii (CT: Caput laterale m. gastrocnemii
et M. soleus); *13* V. saphena parva; *14* Tibia; *15* Lig. collaterale tibiale; *16* Pes anserinus superficialis (Tendines mm. sartorius, gracilis, semitendinosus); *17* V. saphena magna; *18* A. tibialis anterior; *19* N. tibialis et Vasa tibialia posteriora; *20* Caput mediale
m. gastrocnemii; *21* Fascia cruris

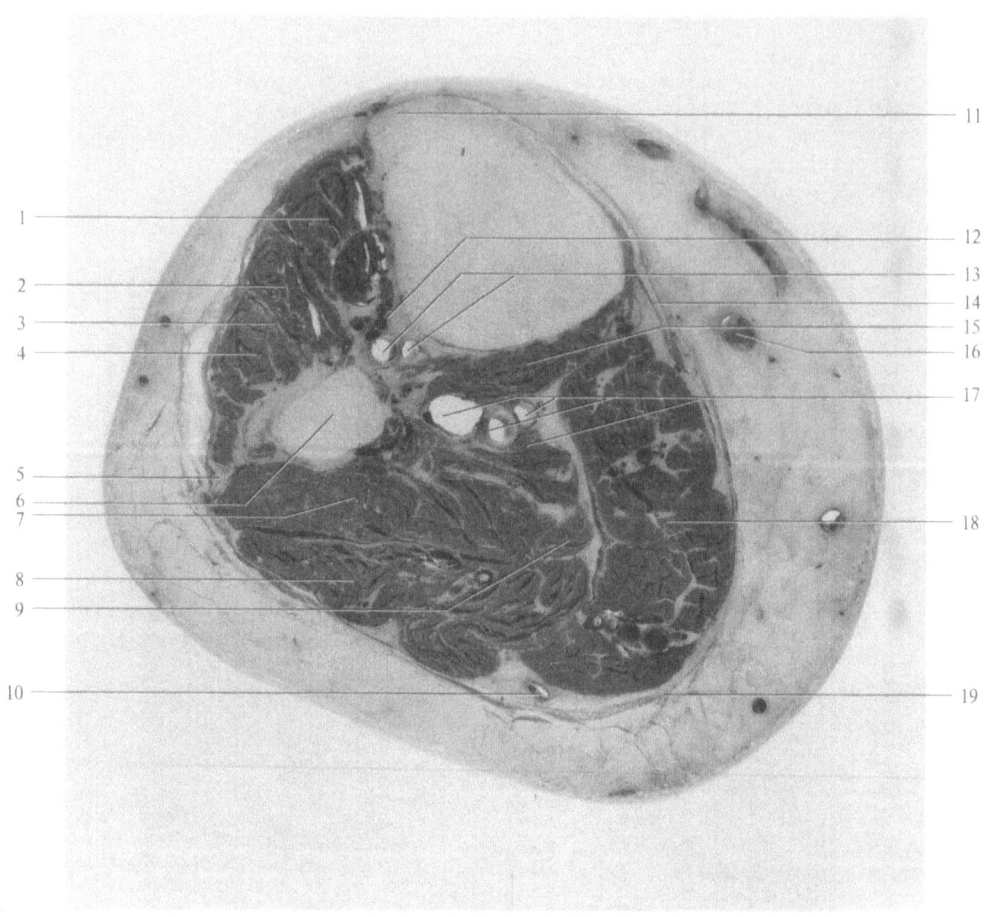

1 —————————
2 —————————
3 —————————
4 —————————
5 —————————
6 —————————
7 —————————
8 —————————
9 —————————
10 —————————

—————————— 11
—————————— 12
—————————— 13
—————————— 14
—————————— 15
—————————— 16
—————————— 17
—————————— 18
—————————— 19

Plate 29

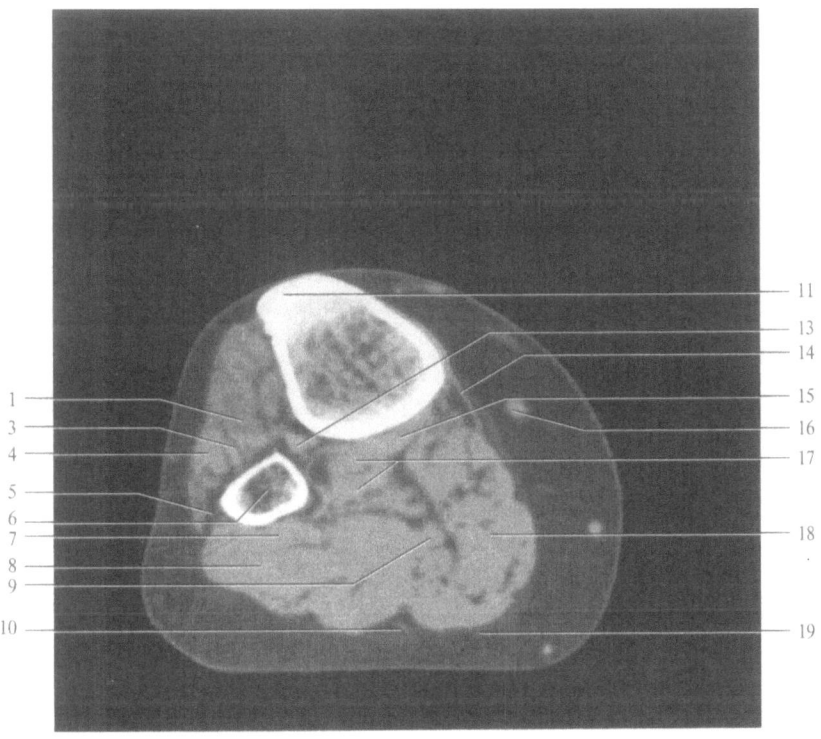

1 M. tibialis anterior (CT: Mm. tibialis anterior et extensor digitorum longus); *2* M. extensor digitorum longus; *3* Septum intermusculare cruris anterius; *4* M. peroneus longus; *5* Septum intermusculare cruris posterius; *6* Corpus fibulae; *7* M. soleus; *8* Caput laterale m. gastrocnemii; *9* M. plantaris; *10* V. saphena parva; *11* Tuberositas tibiae; *12* Membrana interossea; *13* Vasa tibialia anteriora (CT: Vasa tibialia anteriora et Membrana interossea); *14* Pes anserinus superficialis (Tendines mm. sartorius, gracilis, semitendinosus); *15* M. popliteus; *16* V. saphena magna; *17* N. tibialis et Vasa tibialia posteriora; *18* Caput mediale m. gastrocnemii; *19* Fascia cruris

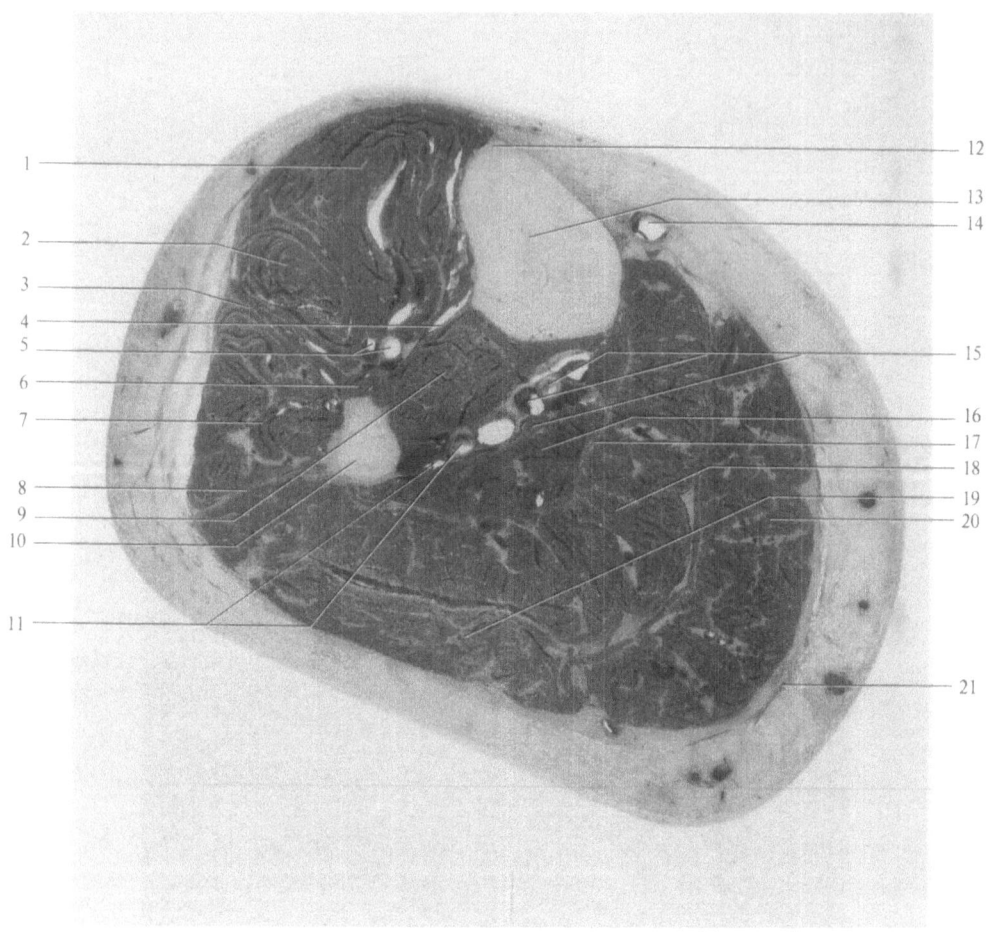

1
2
3
4
5
6
7
8
9
10
11

12
13
14

15

16
17
18
19
20

21

66

Plate 30

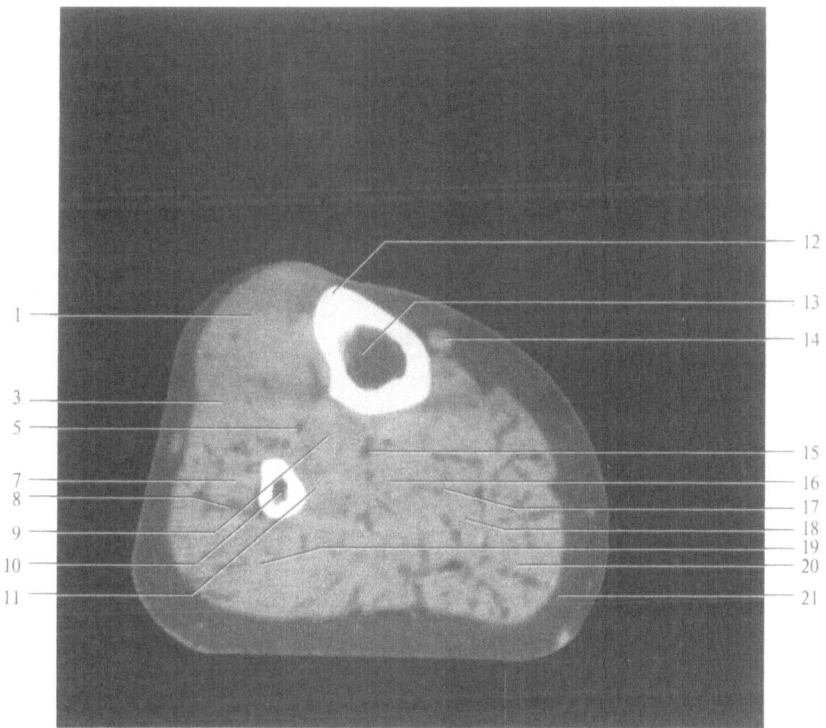

1 M. tibialis anterior (CT: Mm. tibialis anterior et extensor digitorum longus);   *2* M. extensor digitorum longus;   *3* Septum intermusculare cruris anterius;   *4* Membrana interossea;   *5* Vasa tibialia anteriora (CT: Vasa tibialia anteriora et Membrana interossea);   *6* M. extensor hallucis longus;   *7* M. peroneus longus;   *8* Septum intermusculare cruris posterius;   *9* M. tibialis posterior;   *10* Corpus fibulae;   *11* Vasa peronea;   *12* Margo anterior tibiae;   *13* Corpus tibiae;   *14* V. saphena magna;   *15* N. tibialis et Vasa tibialia posteriora;   *16* M. flexor hallucis longus;   *17* Fascia cruris, Lamina profunda;   *18* M. soleus;   *19* Caput laterale m. gastrocnemii;   *20* Caput mediale m. gastrocnemii;   *21* Fascia cruris, Lamina superficialis

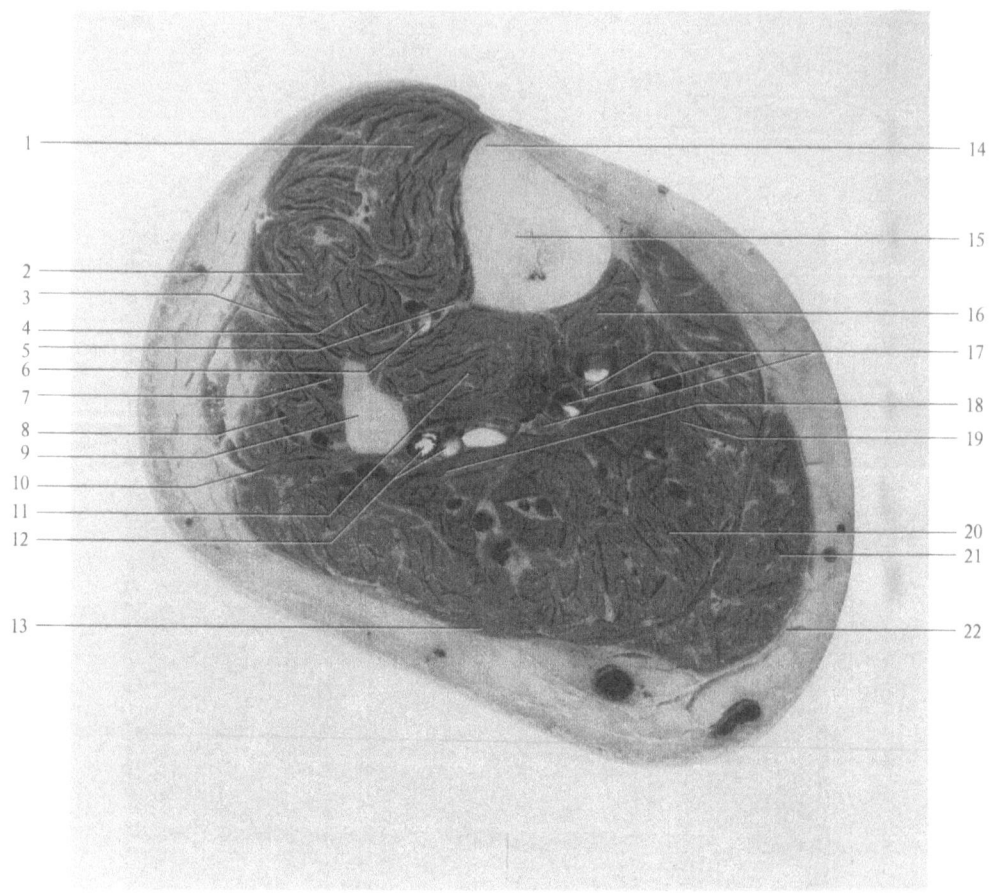

1
2
3
4
5
6
7
8
9
10
11
12
13

14
15
16
17
18
19
20
21
22

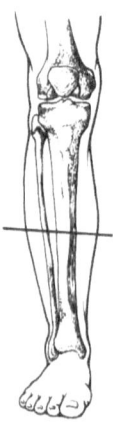

68

Plate 31

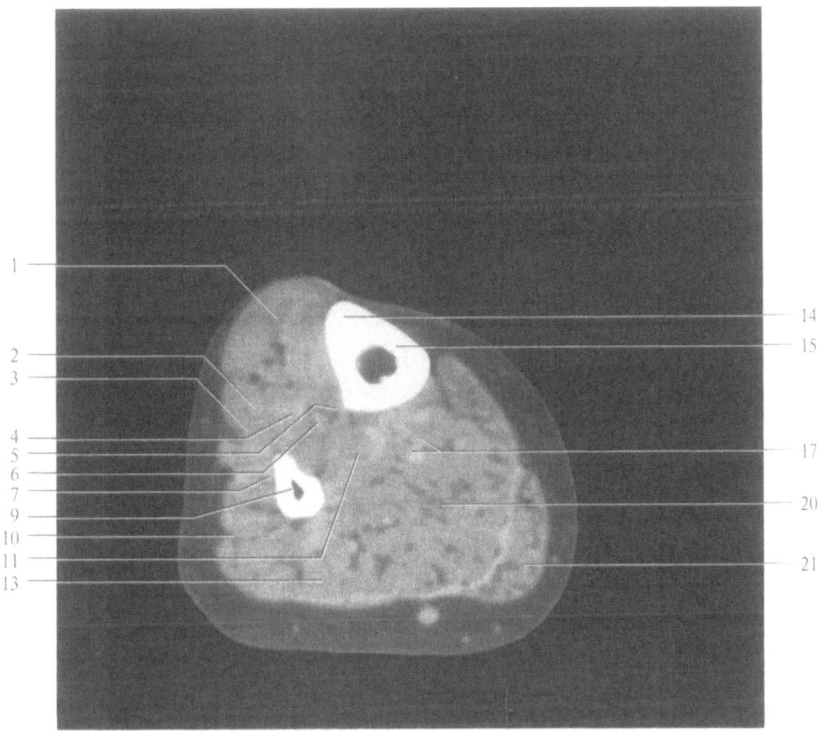

1 M. tibialis anterior; *2* M. extensor digitorum longus; *3* Septum intermusculare cruris anterius; *4* M. extensor hallucis longus; *5* Membrana interossea; *6* N. peroneus profundus et Vasa tibialia anteriora; *7* M. peroneus brevis; *8* M. peroneus longus; *9* Corpus fibulae; *10* Septum intermusculare cruris posterius; *11* M. tibialis posterior; *12* Vasa peronea; *13* Caput laterale m. gastrocnemii; *14* Margo anterior tibiae; *15* Corpus tibiae; *16* M. flexor digitorum longus; *17* N. tibialis et Vasa tibialia posteriora; *18* M. flexor hallucis longus; *19* Fascia cruris, Lamina profunda; *20* M. soleus; *21* Caput mediale m. gastrocnemii; *22* Fascia cruris, Lamina superficialis

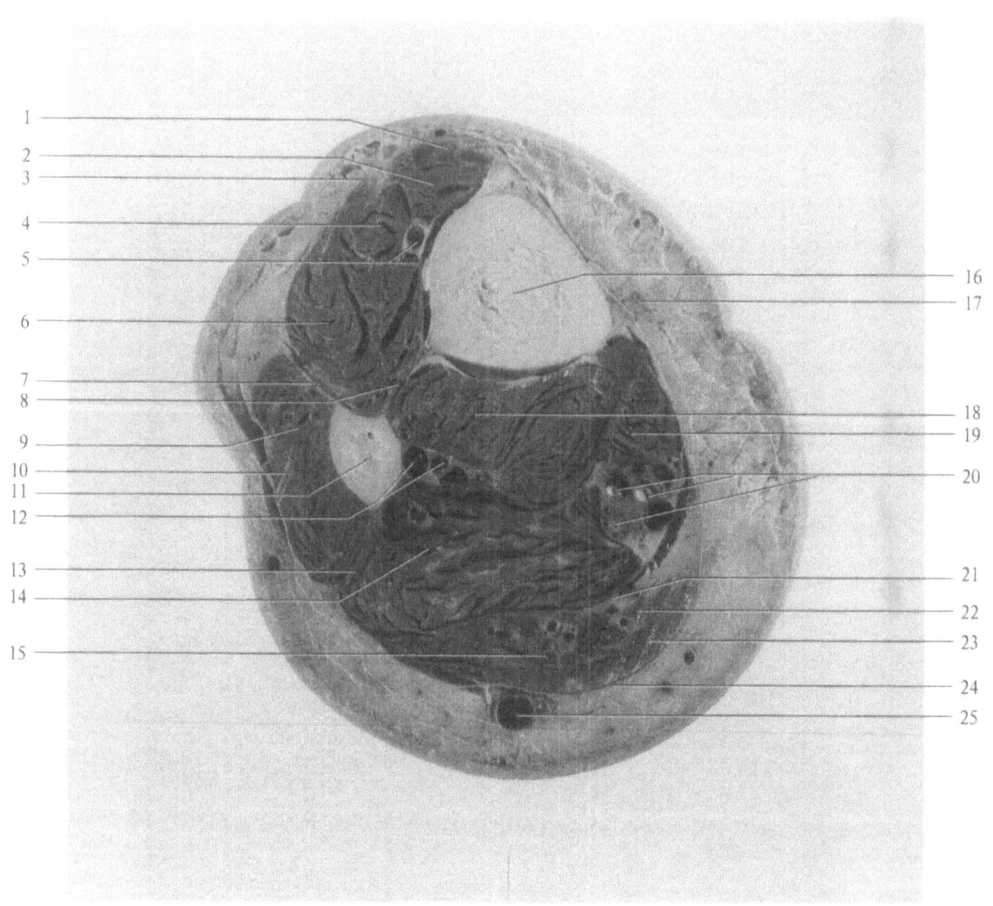

1
2
3

4
5

6

7
8

9
10
11
12

13
14

15

16
17

18
19

20

21
22
23

24
25

70

Plate 32

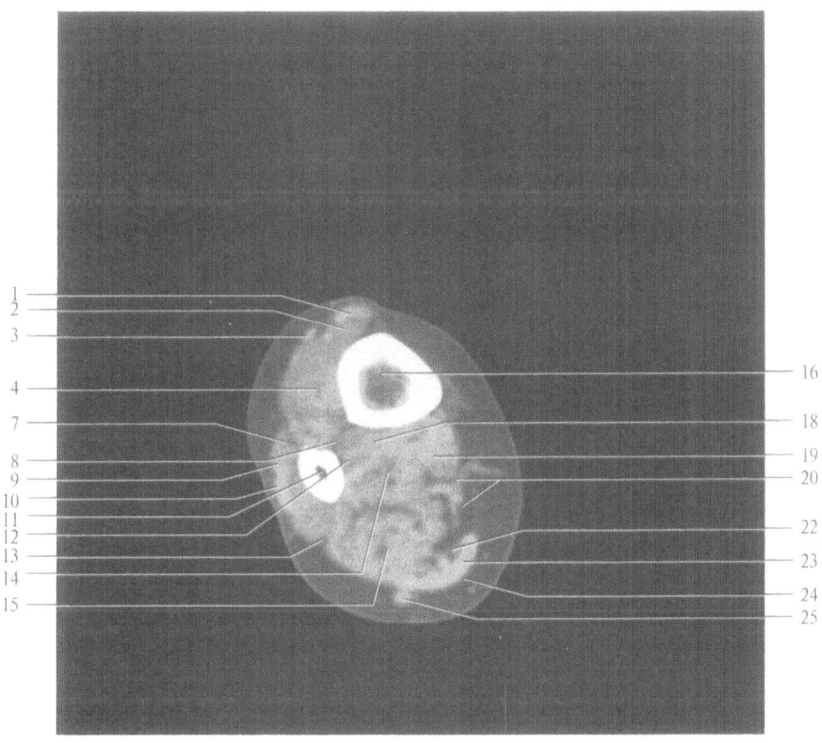

1 Tendo m. tibialis anterioris; *2* M. tibialis anterior; *3* Retinaculum mm. extensorum superius; *4* M. extensor hallucis longus (CT: Mm. extensores hallucis longus et digitorum longus); *5* N. peroneus profundus et Vasa tibialia anteriora; *6* M. extensor digitorum longus; *7* Septum intermusculare cruris anterius; *8* Membrana interossea; *9* M. peroneus brevis; *10* Tendo m. peronei longi; *11* Corpus fibulae; *12* Vasa peronea; *13* Septum intermusculare cruris posterius; *14* M. flexor hallucis longus; *15* M. soleus; *16* Corpus tibiae; *17* N. saphenus; *18* M. tibialis posterior; *19* M. flexor digitorum longus; *20* N. tibialis et Vasa tibialia posteriora; *21* Fascia cruris, Lamina profunda; *22* Tendo m. plantaris; *23* Tendo m. gastrocnemii; *24* Fascia cruris, Lamina superficialis; *25* V. saphena parva

71

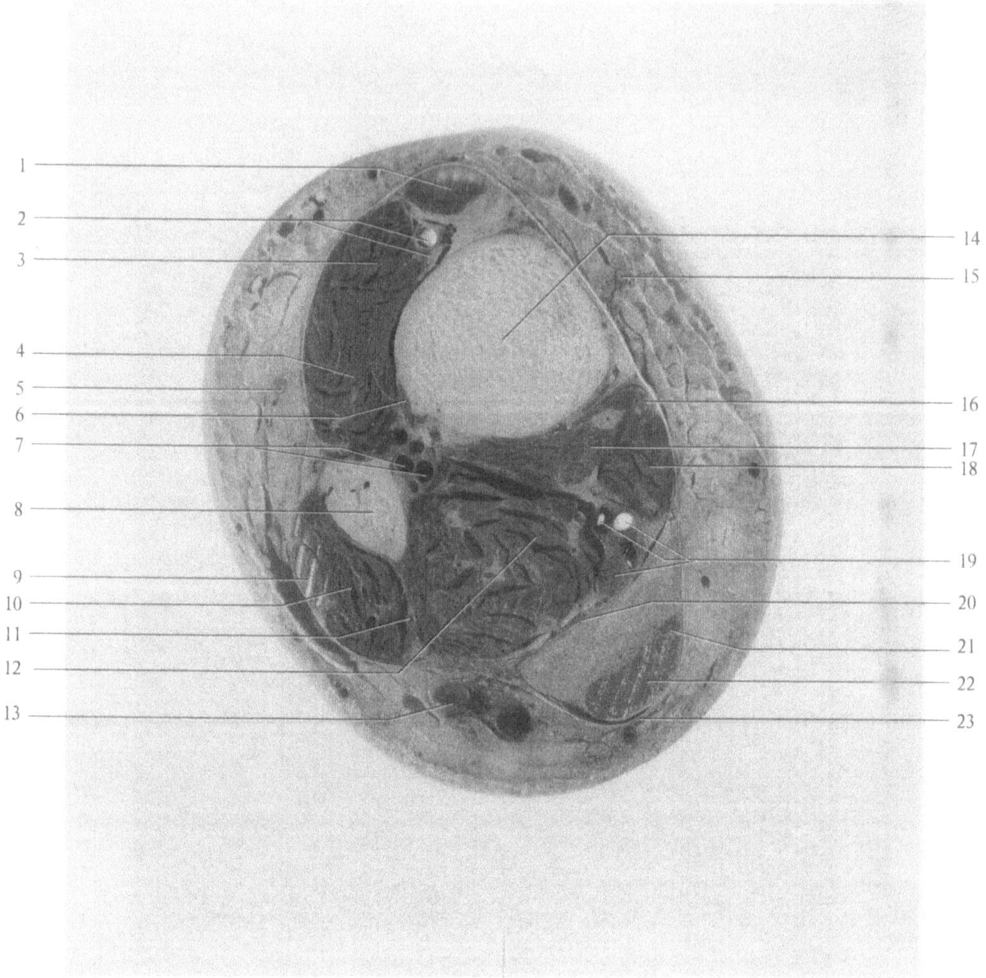

1

2

3

4

5

6

7

8

9

10

11

12

13

14

15

16

17

18

19

20

21

22

23

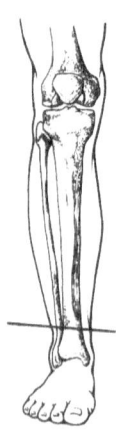

72

Plate 33

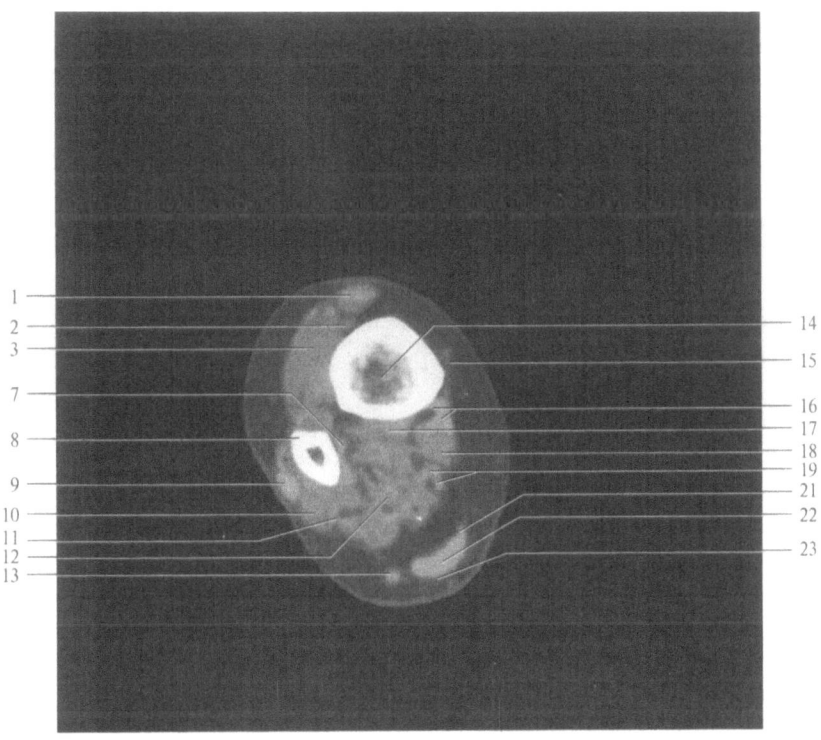

1 Tendo m. tibialis anterioris; 2 N. peroneus profundus et Vasa tibialia anteriora;
3 M. extensor hallucis longus (CT: Mm. extensores hallucis longus et digitorum longus);
4 M. extensor digitorum longus; 5 N. peroneus superficialis; 6 Membrana interossea;
7 Vasa peronea (CT: Vasa peronea et Membrana interossea); 8 Corpus fibulae; 9 Tendo
m. peronei longi; 10 M. peroneus brevis; 11 Septum intermusculare cruris posterius;
12 M. flexor hallucis longus; 13 V. saphena parva; 14 Corpus tibiae; 15 N. saphenus;
16 Tendo m. tibialis posterioris; 17 M. tibialis posterior; 18 M. flexor digitorum longus;
19 N. tibialis et Vasa tibialia posteriora; 20 Fascia cruris, Lamina profunda; 21 Tendo
   m. plantaris; 22 Tendo calcaneus (Achillis); 23 Fascia cruris, Lamina superficialis

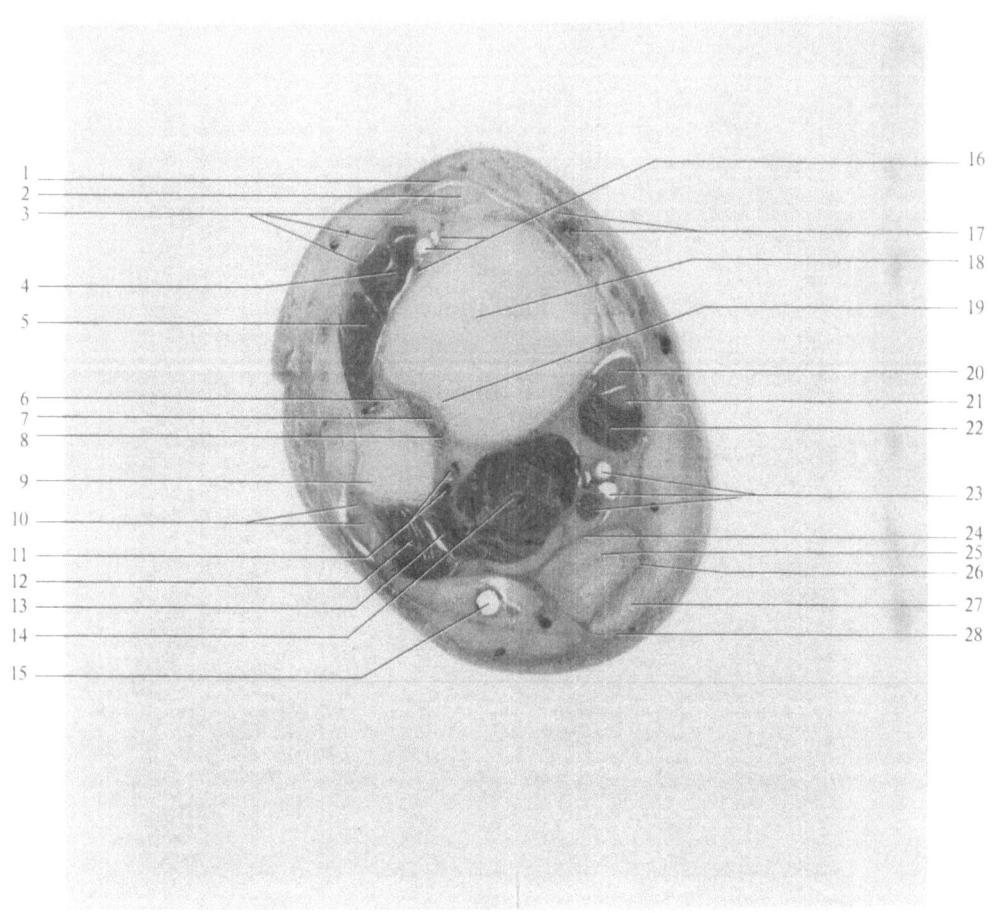

1
2
3

4

5

6
7
8

9

10
11
12
13
14

15

16

17
18

19

20
21
22

23

24
25
26
27
28

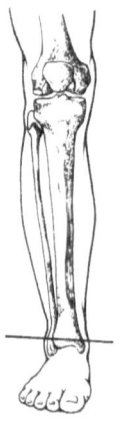

Plate 34

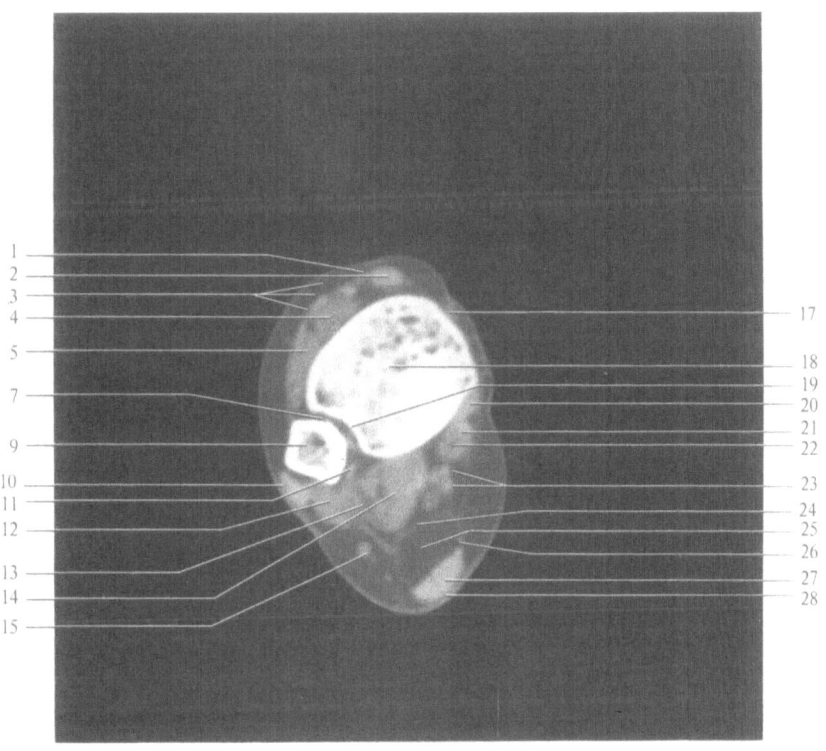

1 Retinaculum mm. extensorum superius; *2* Tendo m. tibialis anterioris; *3* Tendines mm. extensorum hallucis longus et digitorum longus; *4* M. extensor hallucis longus; *5* M. extensor digitorum longus; *6* Lig. tibiofibulare anterius; *7* Syndesmosis tibiofibularis; *8* Lig. tibiofibulare posterius; *9* Malleolus lateralis; *10* Tendines mm. peroneorum brevis et longus; *11* Vasa peronea; *12* M. peroneus brevis; *13* Septum intermusculare cruris posterius; *14* M. flexor hallucis longus; *15* V. saphena parva; *16* N. peroneus profundus, Vasa tibialia anteriora; *17* N. saphenus, V. saphena magna; *18* Tibia; *19* Incisura fibularis tibiae; *20* Tendo m. tibialis posterioris; *21* Tendo m. flexoris digitorum longi; *22* M. flexor digitorum longus; *23* N. tibialis et Vasa tibialia posteriora; *24* Fascia cruris, Lamina profunda; *25* Corpus adiposum subachilleum; *26* Tendo m. plantaris; *27* Tendo calcaneus (Achillis); *28* Fascia cruris, Lamina superficialis

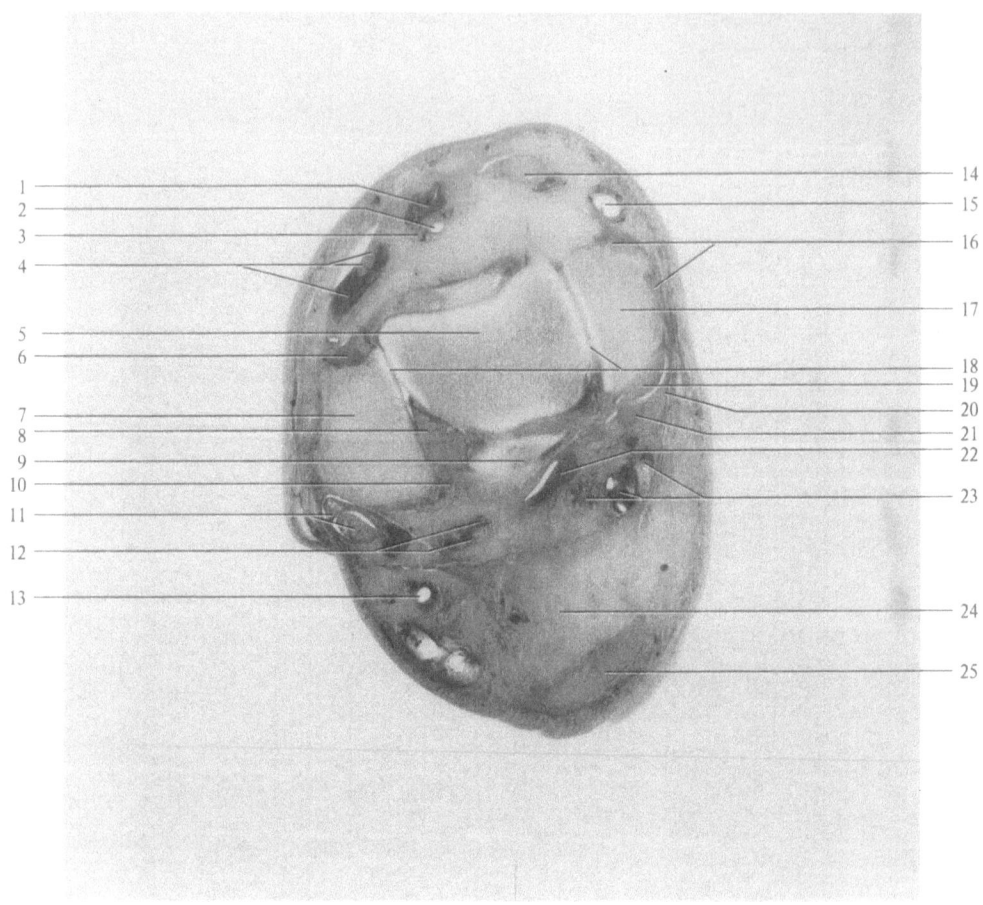

1
2
3
4

5
6

7
8
9
10
11

12

13

14
15
16

17

18
19
20
21
22

23

24

25

Plate 35

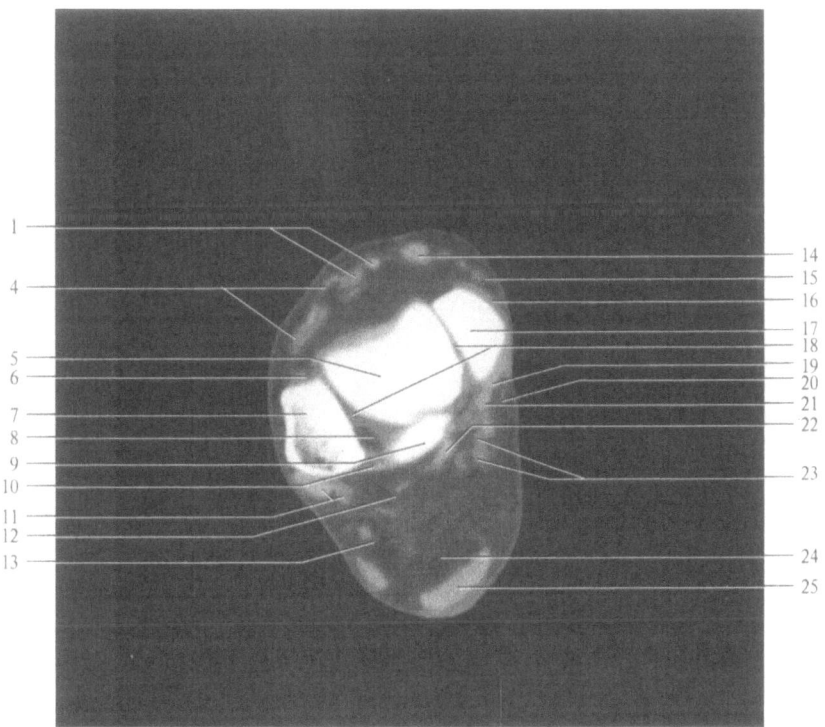

1 M. extensor hallucis longus; *2* A. dorsalis pedis; *3* N. peroneus profundus; *4* M. extensor digitorum longus et Tendines; *5* Trochlea tali; *6* Lig. tibiofibulare anterius; *7* Malleolus lateralis; *8* Capsula articularis; *9* Tibia; *10* Lig. tibiofibulare posterius; *11* Tendines mm. peroneorum brevis et longus; *12* Vasa peronea; *13* V. saphena parva; *14* Tendo m. tibialis anterioris, *15* V. saphena magna; *16* Llg. deltoideum; *17* Malleolus medialis; *18* Articulatio talocruralis; *19* Tendo m. tibialis posterioris; *20* Retinaculum mm. flexorum; *21* Tendo m. flexoris digitorum longi; *22* Tendo m. flexoris hallucis longi; *23* N. tibialis et Vasa tibialia posteriora; *24* Corpus adiposum subachilleum; *25* Tendo calcaneus (Achillis)

77

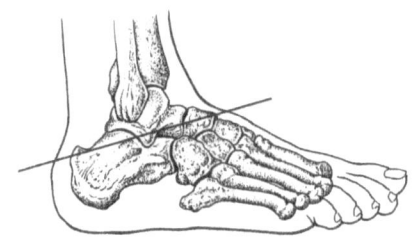

Plate 36

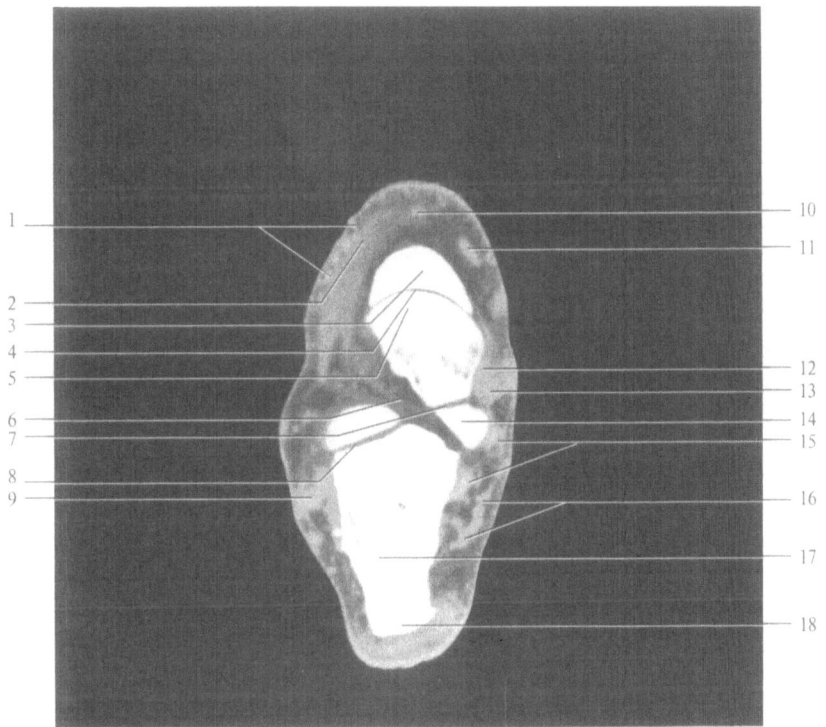

1 Tendines m. extensoris digitorum longi; *2* Mm. extensores hallucis brevis et digitorum brevis; *3* Os naviculare; *4* Articulatio talonavicularis (Pars anterior articulationis talocalcaneonavicularis); *5* Caput tali; *6* Sinus tarsi, Lig. talocalcaneare interosseum; *7* Articulatio talocalcanearis (Pars posterior articulationis talocalcaneonavicularis); *8* Articulatio subtalaris; *9* Tendines mm. peroneorum; *10* Tendo m. extensoris hallucis longi; *11* Tendo m. tibialis anterioris; *12* Lig. calcaneonaviculare plantare; *13* Tendo m. tibialis posterioris; *14* Sustentaculum tali; *15* Tendines mm. flexorum digitorum longus et hallucis longus; *16* Aa. plantares medialis et lateralis, Nn. plantares medialis et lateralis; *17* Calcaneus; *18* Tuber calcanei

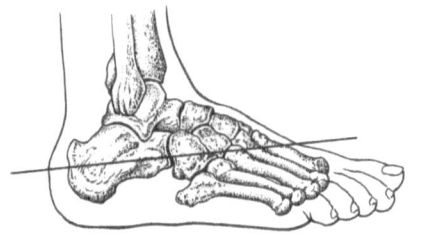

Plate 37

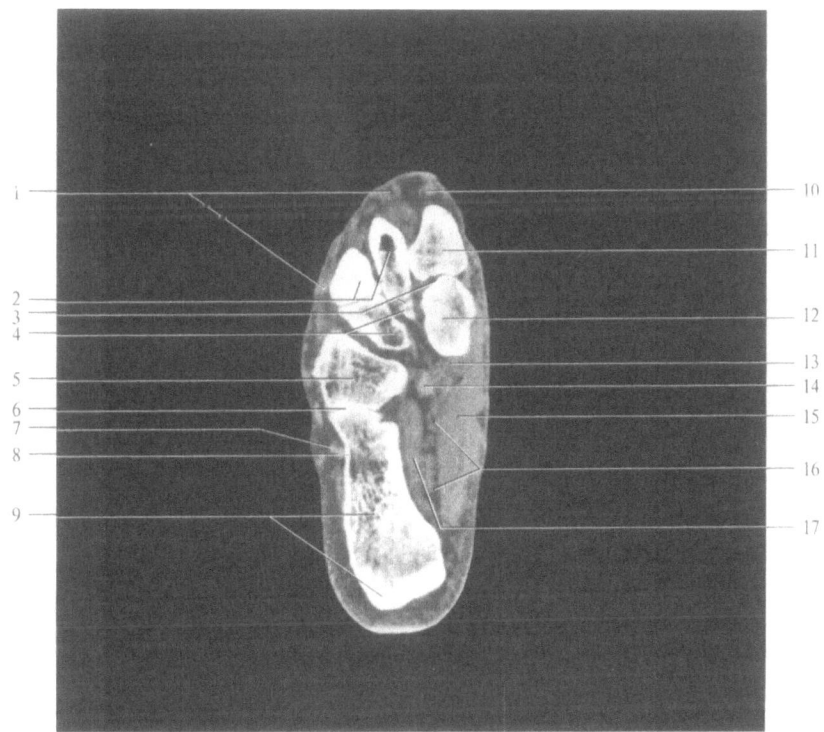

1 Tendines m. extensoris digitorum longi, Mm. extensores hallucis brevis et digitorum brevis; 2 Ossa metatarsalia II, III; 3 Articulationes tarsometatarsales I-III (Lisfranc); 4 Ossa cuneiformia intermedium et laterale; 5 Os cuboideum; 6 Articulatio calcaneocuboidea; 7 Tendo m. peronei brevis; 8 Tendo m. peronei longi; 9 Calcaneus, Tuber calcanei; 10 Tendo m. extensoris hallucis longi; 11 Basis ossis metatarsalis I; 12 Os cuneiforme mediale; 13 Insertio m. tibialis posterioris; 14 Tendines mm. flexorum digitorum longus et hallucis longus; 15 Mm. abductor hallucis et flexor hallucis brevis; 16 Aa. plantares medialis et lateralis, Nn. plantares medialis et lateralis; 17 M. quadratus plantae

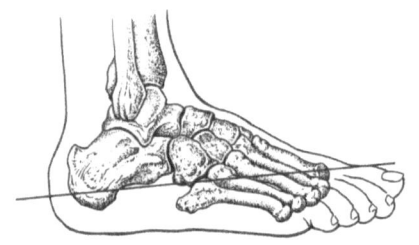

Plate 38

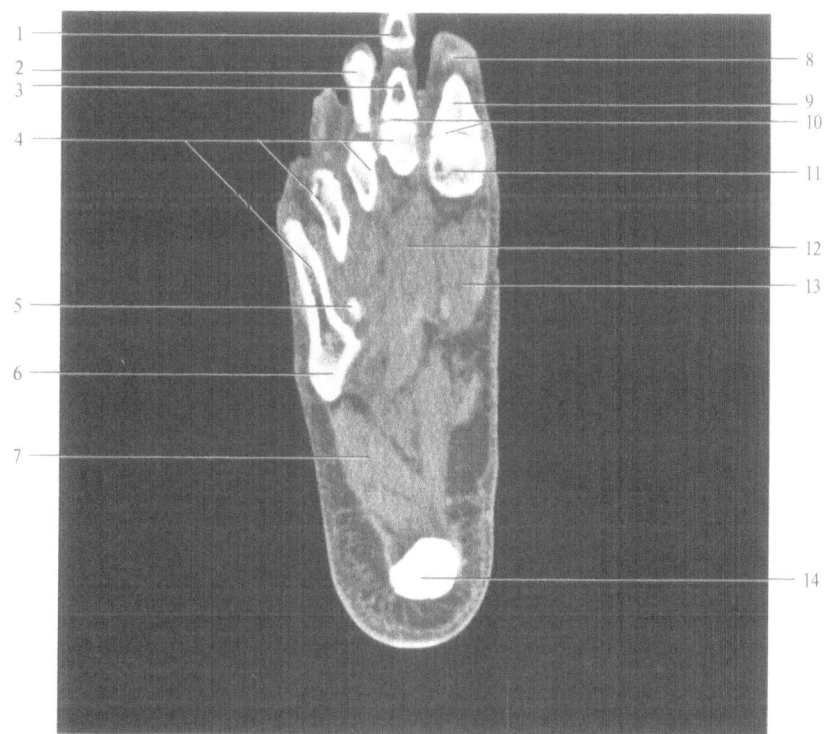

1 Phalanx media digiti II; 2 Phalanx proximalis digiti III; 3 Phalanx proximalis digiti II; 4 Ossa metatarsalia II–V; 5 Basis ossis metatarsalis IV; 6 Tuberositas ossis metatarsalis V.; 7 M. abductor digiti minimi; 8 Tendo m. extensoris hallucis longi; 9 Phalanx proximalis hallucis; 10 Articulationes metatarsophalangeales I et II; 11 Caput ossis metatarsalis I; 12 Caput obliquum m. adductoris hallucis; 13 M. flexor digitorum brevis; 14 Tuber calcanei

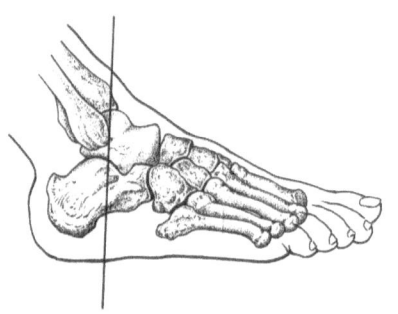

Plate 39

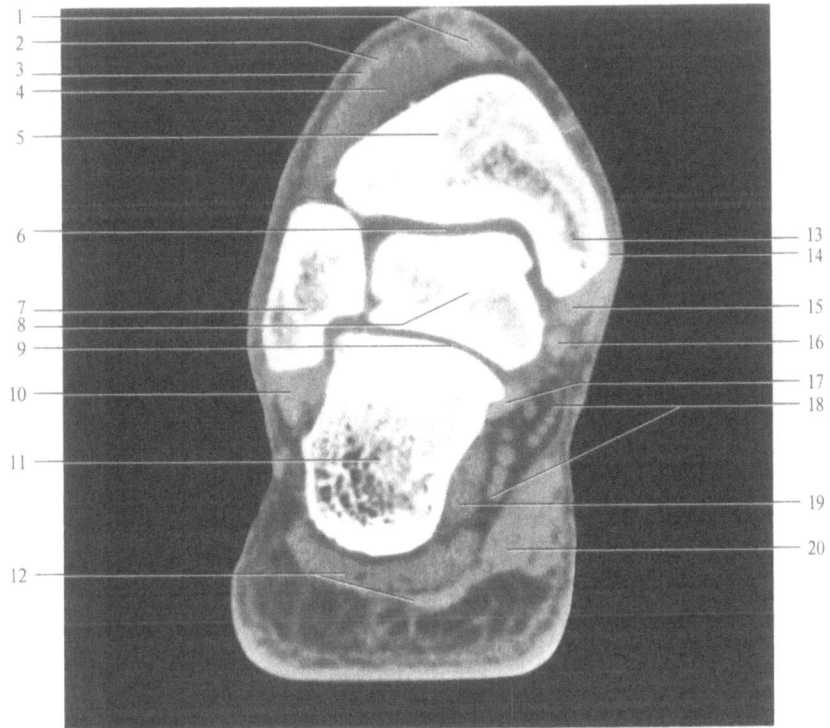

1 Tendo m. tibialis anterioris; *2* Tendo m. extensoris hallucis longi; *3* Tendo m. extensoris digitorum longi; *4* Mm. extensores hallucis longus et digitorum longus; *5* Tibia; *6* Articulatio talocruralis; *7* Fibula; *8* Talus; *9* Articulatio subtalaris; *10* Tendines mm. peroneorum longus et brevis; *11* Calcaneus; *12* M. flexor digitorum brevis et Aponeurosis plantaris; *13* Malleolus medialis; *14* Lig. deltoideum; *15* Tendo m. tibialis posterioris; *16* Tendo m. flexoris digitorum longi; *17* Tendo m. flexoris hallucis longi; *18* Vasa tibialia posteriora et Nn. plantares medialis et lateralis; *19* M. quadratus plantae; *20* M. abductor hallucis

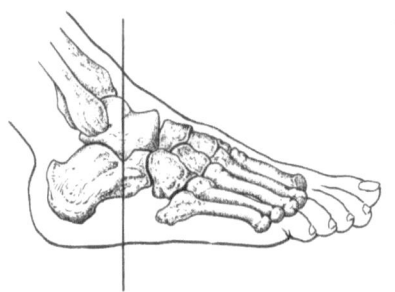

Plate 40

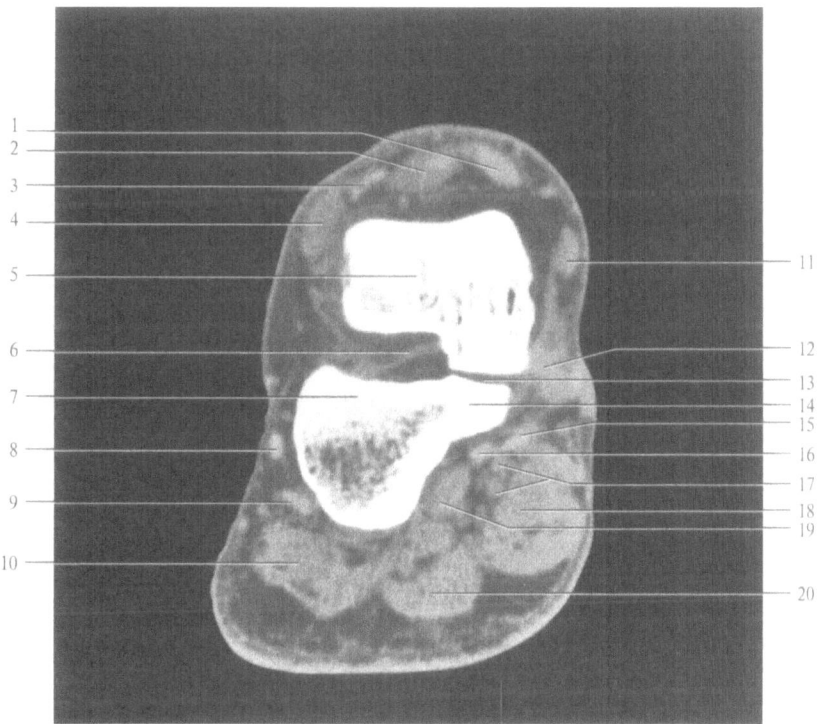

1 Tendo m. extensoris hallucis longi; 2 M. extensor hallucis brevis; 3 Tendo m. extensoris digitorum longi; 4 M. extensor digitorum brevis; 5 Talus; 6 Sinus tarsi, Lig. talocalcaneare interosseum; 7 Calcaneus; 8 Tendo m. peronei brevis; 9 Tendo m. peronei longi; 10 M. abductor digiti minimi; 11 Tendo m. tibialis anterioris; 12 Tendo m. tibialis posterioris et Lig. calcaneonaviculare plantare; 13 Articulatio talocalcaneonavicularis (Pars posterior); 14 Sustentaculum tali; 15 Tendo m. flexoris digitorum longus; 16 Tendo m. flexoris hallucis longus; 17 A., V., N. plantaris medialis; 18 M. abductor hallucis; 19 M. quadratus plantae; 20 M. flexor digitorum brevis et Aponeurosis plantaris

Plate 41

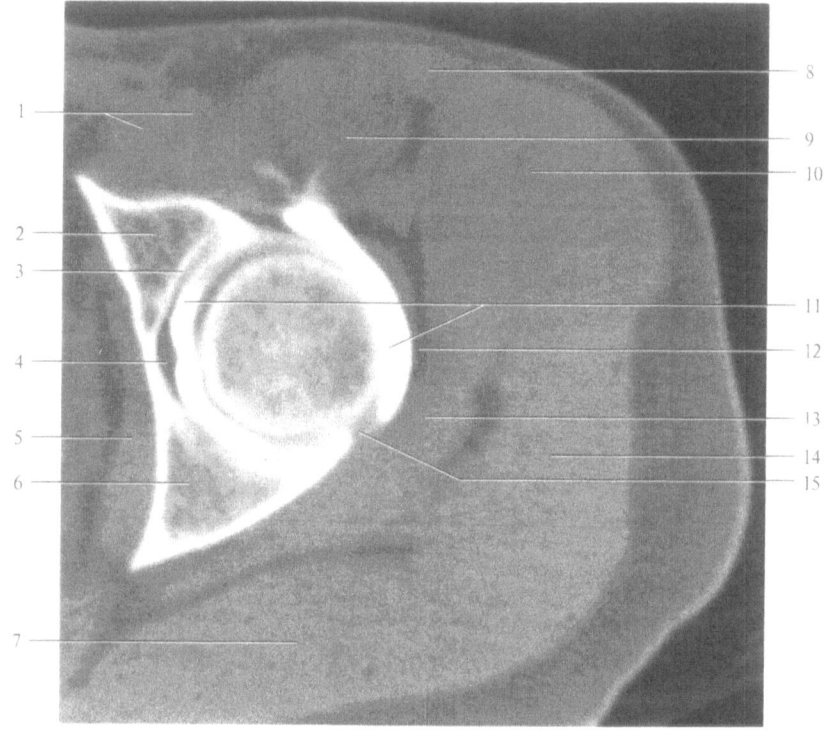

1 Vasa femoralia; *2* Corpus ossis pubis; *3* Facies lunata; *4* Fossa acetabuli; *5* M. obturator internus; *6* Corpus ossis ischii; *7* M. gluteus maximus; *8* M. sartorius; *9* M. iliopsoas; *10* M. tensor fasciae latae; *11* Cavitas articularis; *12* Capsula articularis; *13* M. gluteus minimus; *14* M. gluteus medius; *15* Labrum acetabulare

Plate 42

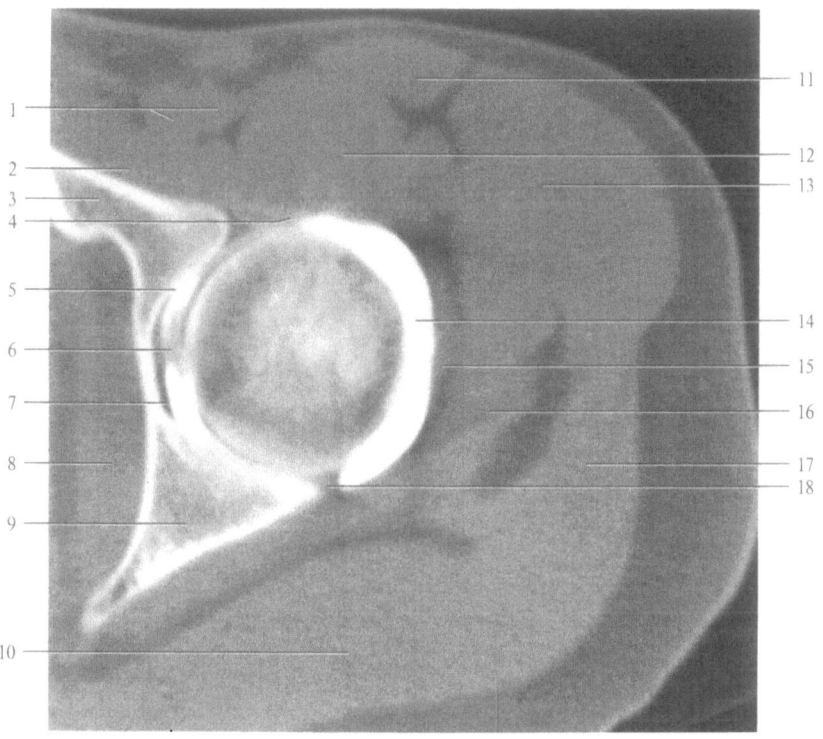

1 Vasa femoralia; *2* M. pectineus; *3* Ramus superior ossis pubis; *4* Lig. pubofemorale, Labrum acetabulare; *5* Facies lunata; *6* Lig. capitis femoris; *7* Fossa acetabuli; *8* M. obturator internus; *9* Corpus ossis ischii; *10* M. gluteus maximus; *11* M. sartorius; *12* M. iliopsoas; *13* M. tensor fasciae latae; *14* Cavitas articularis; *15* Capsula articularis; *16* M. gluteus minimus; *17* M. gluteus medius; *18* Labrum acetabulare

Plate 43

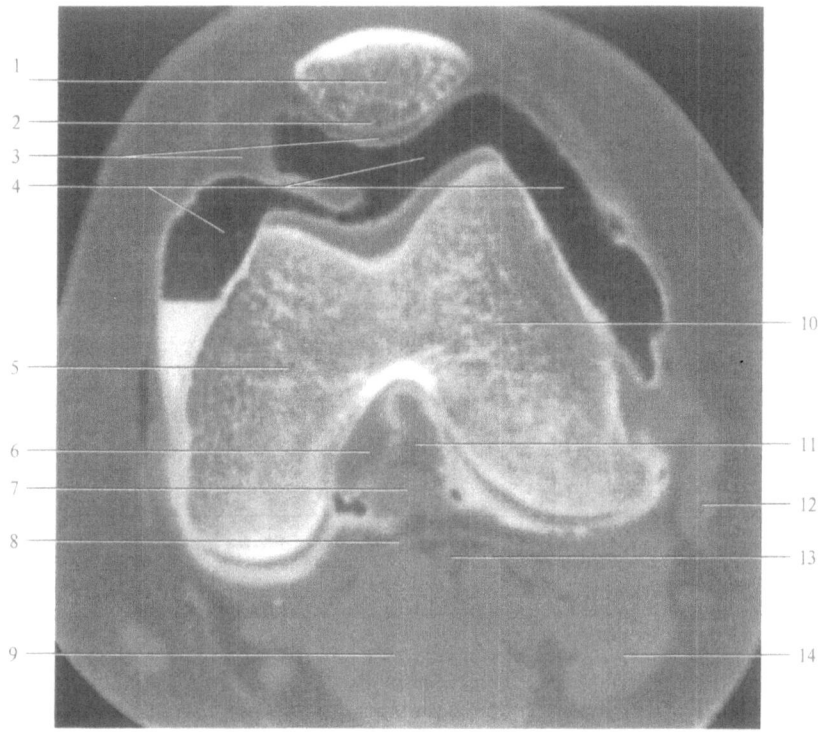

1 Patella; *2* Facies articularis patellae; *3* Corpus adiposum infrapatellare; *4* Cavitas articularis; *5* Condylus medialis ossis femoris; *6* Lig. cruciatum posterius; *7* Lig. menis-cofemorale; *8* Lig. popliteum obliquum; *9* Caput mediale m. gastrocnemii; *10* Condylus lateralis ossis femoris; *11* Lig. cruciatum anterius; *12* M. biceps femoris; *13* Vasa poplitea; *14* Caput laterale m. gastrocnemii

Plate 44

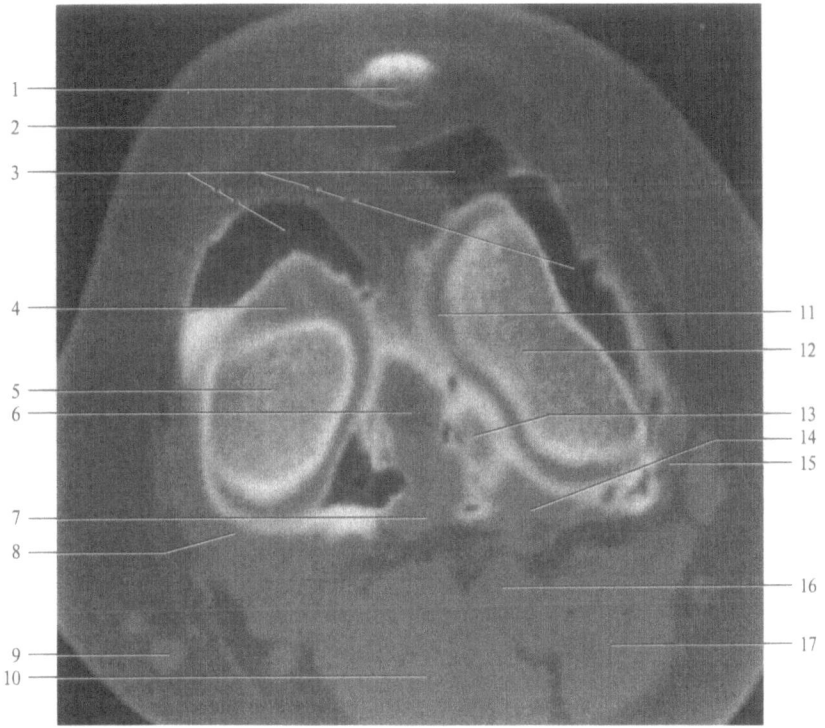

1 Patella; *2* Corpus adiposum infrapatellare; *3* Cavitas articularis; *4* Cartilago condyli medialis ossis femoris; *5* Condylus medialis ossis femoris; *6* Lig. cruciatum anterius; *7* Lig. cruciatum posterius; *8* Meniscus medialis (cornu posterius); *9* V. saphena magna; *10* Caput mediale m. gastrocnemii; *11* Cartilago condyli lateralis ossis femoris; *12* Condylus lateralis ossis femoris; *13* Tuberculum intercondylare laterale tibiae; *14* Meniscus lateralis (cornu posterius); *15* Lig. collaterale fibulare; *16* M. plantaris; *17* Caput laterale m. gastrocnemii

Plate 45

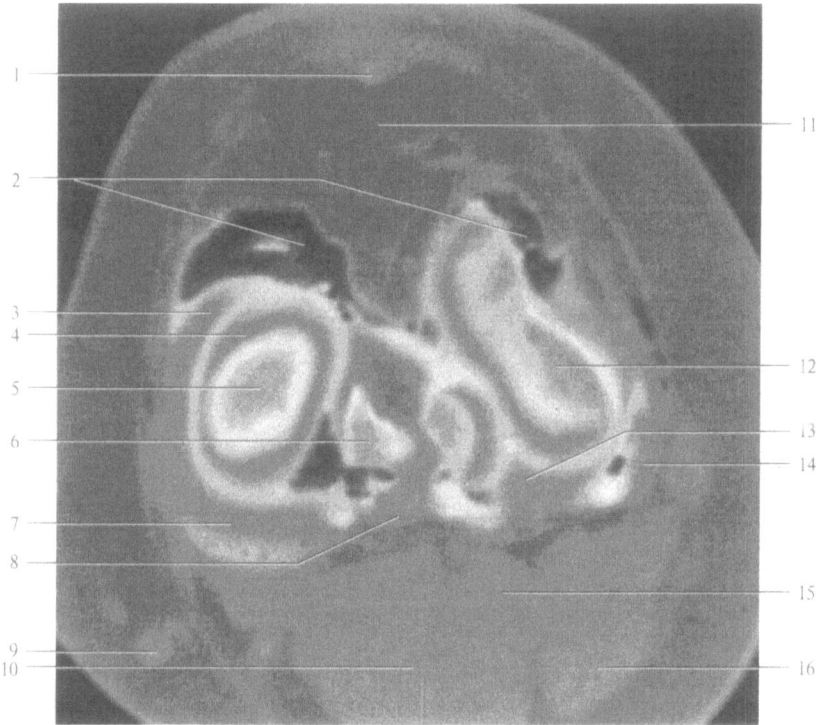

1 Patella; *2* Cavitas articularis; *3* Meniscus medialis (cornu anterius); *4* Cartilago condyli medialis ossis femoris; *5* Condylus medialis ossis femoris; *6* Tuberculum intercondylare mediale tibiae; *7* Meniscus medialis (cornu posterius); *8* Lig. meniscofemorale posterius; *9* V. saphena magna; *10* Caput mediale m. gastrocnemii; *11* Corpus adiposum infrapatellare; *12* Condylus lateralis ossis femoris; *13* Meniscus lateralis (cornu posterius); *14* Lig. collaterale fibulare; *15* M. plantaris; *16* Caput laterale m. gastrocnemii

Plate 46

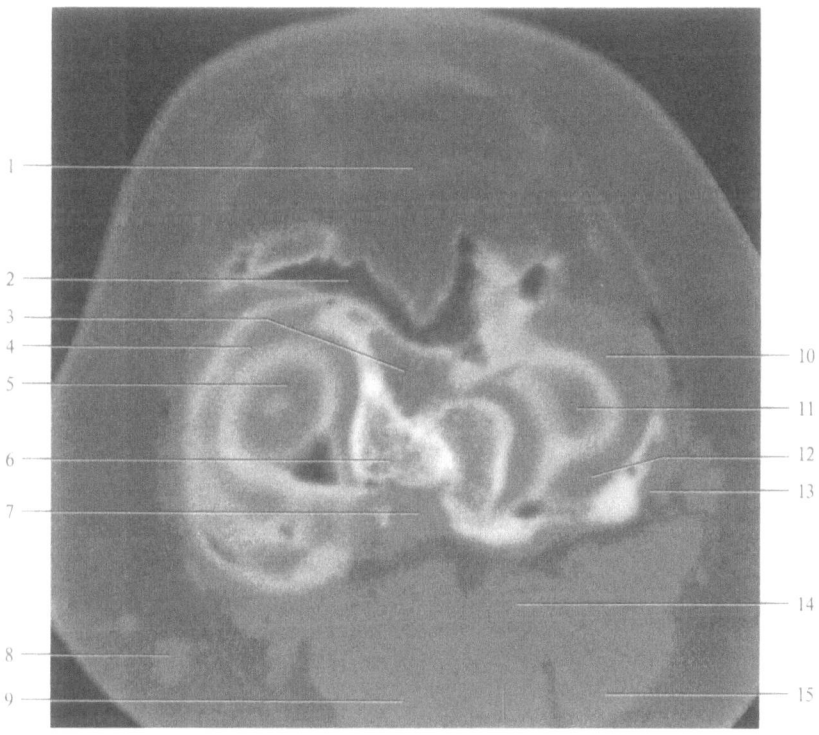

1 Corpus adiposum infrapatellare; *2* Cavitas articularis; *3* Cartilago articularis tibiae; *4* Meniscus medialis; *5* Cartilago condyli medialis ossis femoris; *6* Tuberculum inter-condylare mediale tibiae; *7* Lig. cruciatum posterius; *8* V. saphena magna; *9* Caput mediale m. gastrocnemii; *10* Meniscus lateralis (cornu anterius); *11* Cartilago condyli lateralis ossis femoris; *12* Meniscus lateralis (cornu posterius); *13* Lig. collaterale fibu-lare; *14* M. plantaris; *15* Caput laterale m. gastrocnemii

Plate 47

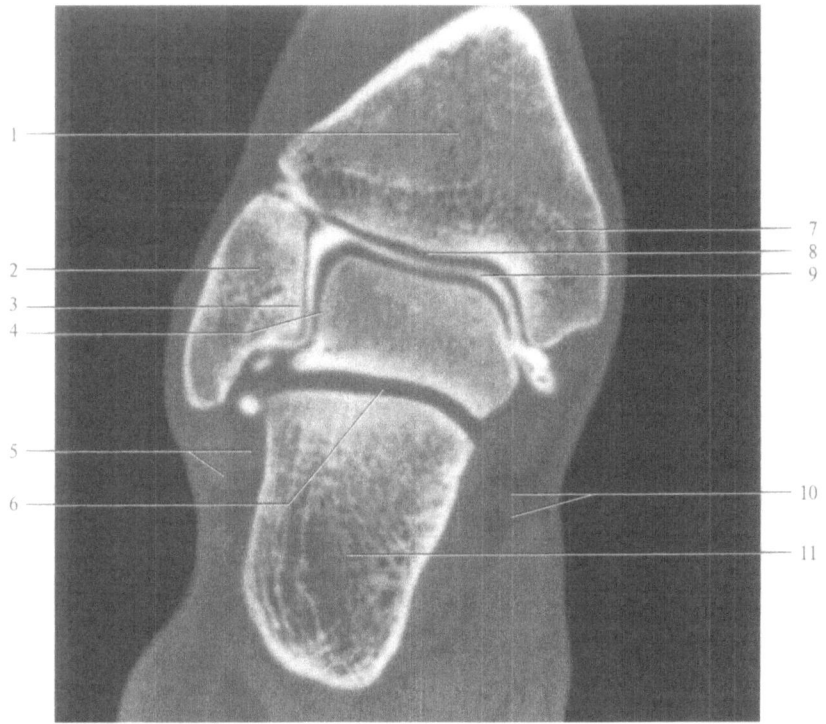

1 Tibia; *2* Fibula; *3* Cartilago articularis ossis fibulae; *4* Cartilago articularis ossis tali; *5* Tendines mm. peroneorum longus et brevis; *6* Articulatio subtalaris; *7* Malleolus medialis; *8* Cartilago articularis ossis tibiae; *9* Articulatio talocruralis; *10* Vasa tibialia posteriora et nn. plantares medialis et lateralis; *11* Calcaneus

Plate 48

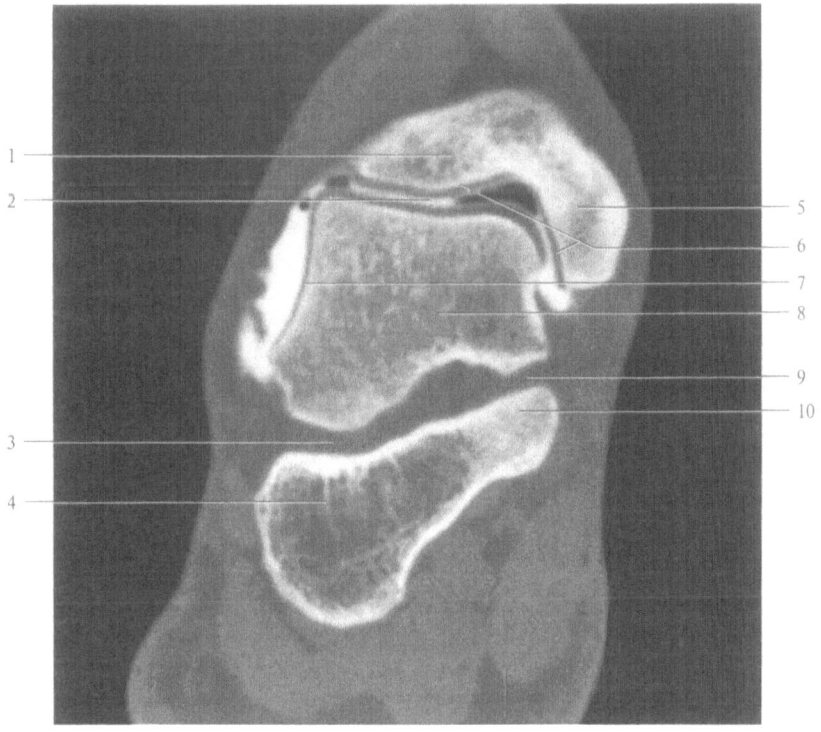

1 Tibia; *2* Articulatio talocruralis; *3* Articulatio subtalaris; *4* Calcaneus; *5* Malleolus medialis; *6* Cartilago articularis ossis tibiae; *7* Cartilago articularis ossis tali; *8* Talus; *9* Articulatio talocalcanearis (Pars posterior articulationis talocalcaneonavicularis); *10* Sustentaculum tali

Measurements of the Lower Extremity

In orthopedics, knowledge of angles and distances is essential for the assessment of normality, the distinction of pathologic from normal conditions, the quantification of pathologic states, and preoperative planning. Although the various angles and distances measured in the lower extremity do not replace clinical examinations, they help the surgeon to reach a diagnosis. In the hip joint, the center collum-diaphysis (CCD) angle, the anteversion (AT) angle, and the center-edge (CE) angle are often measured. In the knee joint, the distance on the axial plane between the groove of the facies patellaris femoris and the tuberositas tibiae (FPG-TT distance) can be measured and may help to explain patellar pain syndromes. Torsion of the tibia can also be assessed. Finally, length measurements and axis measurements can also be performed with CT.

There is a lot of controversy in the literature concerning the methods to be used in CT measurements, with the greatest discussion always about the definition of the reference points. The methods we propose are simple and effective, but the reader may have a preference for others, and we believe that radiologists should use the method they prefer, as no method is absolutely accurate. It is, however, essential to use the same method in both extremities and for follow-up studies, as different methods yield different results. Further, in practice, CT measurement of the femoral or tibial torsion will generally be more accurate than conventional methods.

The Anteversion Angle

Conventional Measurement

Anteversion is conventionally measured from two radiographs. One is the standard pelvic view (Fig. 1 a) with the patient's legs hanging over the edge of the X-ray table in order to get the line connecting the medial and lateral femoral condyles parallel to the surface of the table. The X-ray beam is centered on the symphysis (Fig. 1 c). The second is the Rippstein-Dunn view (Fig. 1 b), for which the patient lies supine, the

Figure 1 a–e

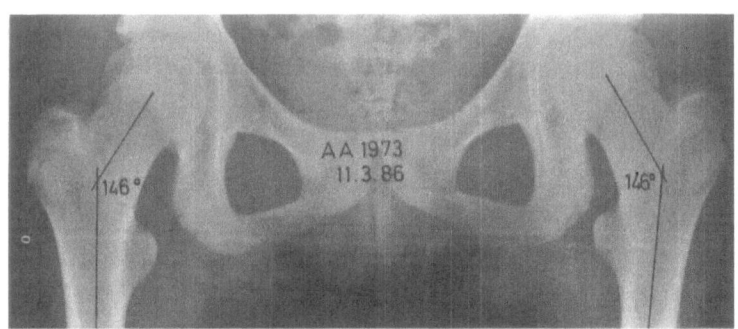

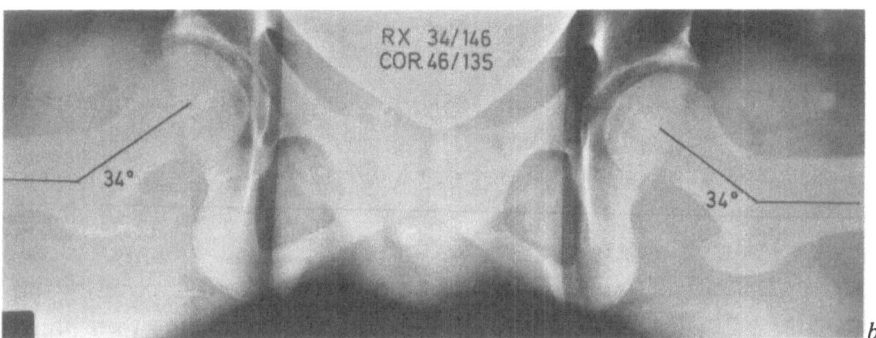

a Measurement of the CCD angle. Projected angle 146° on both sides. The corrected angle, knowing the value of the projected antetorsion, is 135°. This value corresponds to the real CCD angle. *b* Measurement of the AT angle on the Rippstein-Dunn view. The projected angle is 34° on both sides, the corrected AT angle 46° on both sides. This latter value corresponds to the real antetorsion. *c* The standard view of the pelvis with hanging legs. On this view, the line connecting the medial and lateral femoral condyles *(C)* should be parallel to the surface of the table. *Arrow,* X-ray beam. (Redrawn from Hafner and Meuli 1975). *d,e* The Rippstein-Dunn view for measurement of AT. *Arrow,* X-ray beam. (Redrawn from Hafner and Meuli 1975)

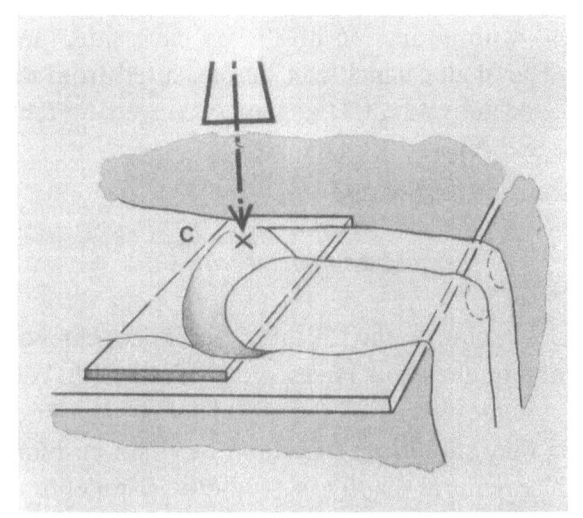

c

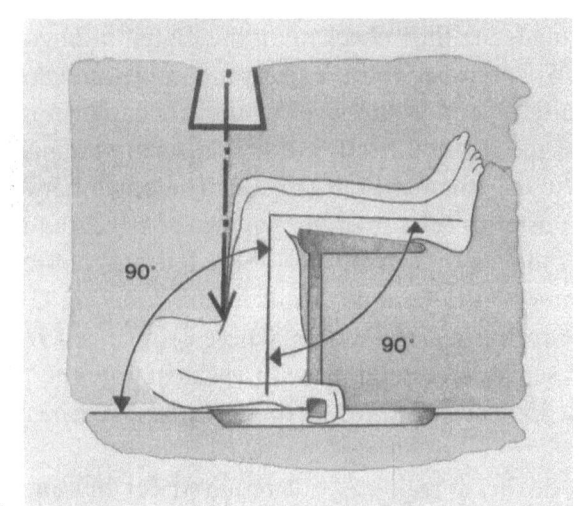

d

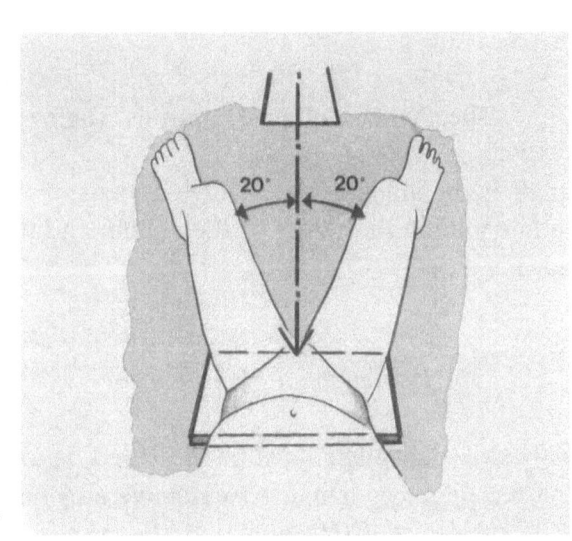

e

hip flexed 90° with abduction of 20° on each side, the tibias parallel (Fig. 1 d, e). The angles which can be measured from these two views are the projected AT and CCD angles. A conversion table is then used to obtain the real AT and CCD angles.

Measurement with CT

The patient lies supine on the CT table with knees extended. In order to ensure immobility, feet and knees are secured with Velcro straps. As there is physiologic valgus of the distal femur, the knees should not touch; rather, they should be separated by foam cushions. Ideally, the long axis of the femoral diaphysis should be parallel to the table plane and to the long axis of the CT table. The position and the level of the scans to be made are then checked on the digital radiograph (scout view or topogram). Three 4-mm-thick scans are made, one at the level of the femoral condyles, one at the level of the trochanter minor, and one at the center of the femoral head. These images are then added (superimposed) by the CT scanner itself (Fig. 2). The angle between a line connecting the posterior aspects of the femoral condyles and a line between the center of the femoral diaphysis (at the level of the trochanter minor) and the center of the femoral head on the resulting composite image corresponds to the real AT angle. There is no need for a conversion table or for separate computing with a calculator: the measurement is precise and does not depend on the accuracy of measurement of the CCD angle.

As high-quality imaging is not required for this measurement, the radiation dose is set at the lowest level.

The Center-Collum-Diaphysis Angle

The CCD angle is the angle between the long axis of the femoral diaphysis and a line joining the center of the collum and the center of the femoral head (Fig. 1 a).

Conventional Measurement

Conventionally, the CCD angle must be measured on a standard pelvic view with the patient's legs hanging (see above and Fig. 1). The value obtained is the projected CCD angle. If the projected AT angle is

Figure 2

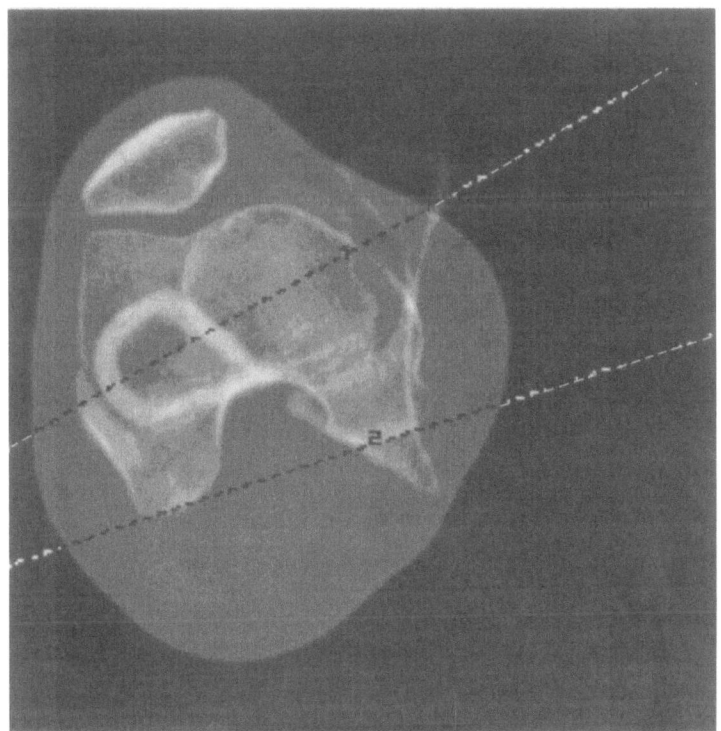

Measurement of anteversion with CT. This image is superimposition of three scans. On the monitor, one line is drawn joining the center of the femoral head and the center of the diaphysis *(1)*. A second line is drawn joining the dorsal borders of the medial and lateral femoral condyles (bicondylar plane) *(2)*. The angle between these lines corresponds to the real anteversion, in this case 12°

known, the value of the real angles is read off from a conversion table. Again, the accuracy of the value obtained for the CCD angle depends on how precisely the AT angle was measured.

Measurement with CT

Although the CCD angle is easily measured on a standard pelvic radiograph, which is generally available, it can also be measured using CT. The patient lies prone on the CT table and the hip is rotated inward to an angle equal to the previously measured real AT angle as measured on CT. The long axis of the femoral diaphysis must be aligned along the centerline of the CT table parallel to its longitudinal axis. In this position it is assumed that the anteversion is compensated and the collum is parallel to the table plane. The digital radiograph (scout view or topogram) is then taken and the CCD angle can be measured on the monitor. Again, precise positioning of the patient is necessary in order to obtain accurate measurements.

The Center-Edge Angle

The CE angle, the angle between the longitudinal axis of the body and a line from the edge of the acetabulum to the center of the femoral head (Fig. 3), gives a rough appreciation of the coverage of the femoral head by the acetabulum (see next chapter for detailed discussion of this topic). The CE angle can be measured on the digital radiograph of the CT but should be measured from a standard pelvic view, which is almost always available, or, even better, from a Dunn view. The resolution of a conventional view is higher than that of the CT digital radiograph, so the measurement is more precise.

The Distance Between the Facies Patellaris Femoris Groove
and the Tuberositas Tibiae

The measurement of the FPG-TT distance may be helpful in understanding femoropatellar dysfunctions. The distance is measured with 30° flexion of the knee, the patient lying supine on the CT table. Although the measurement is easy, the patient must be precisely positioned. The foot should be in a natural position with outward rotation of approximately 15°. The patient should be absolutely motionless, so it

Figure 3

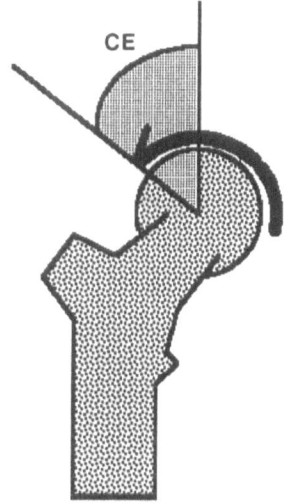

Definition of the CE angle. Normal: > 25° ; pathological: < 16°

Figure 4

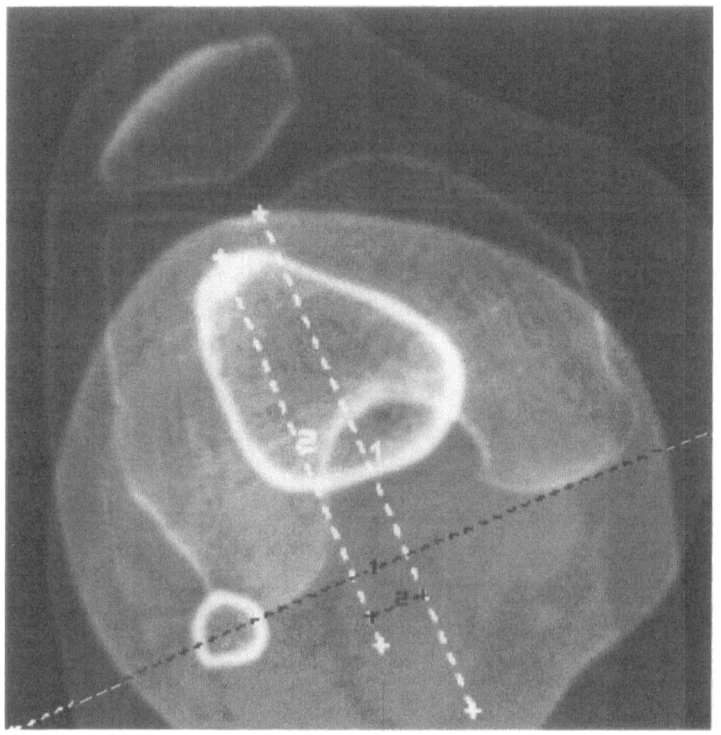

Measurement of the FPG-TT distance. This image is an addition of two scans (see text). A line *(1)* is drawn perpendicular to the bicondylar plane passing through the lowest part of the groove of the facies patellaris. A second line *(2)*, also perpendicular to the bicondylar plane, passes through the tuberositas tibiae at the most ventral level. The distance between line *1* and line *2* is the distance to be measured. In this case it was normal (9 mm)

Figure 5

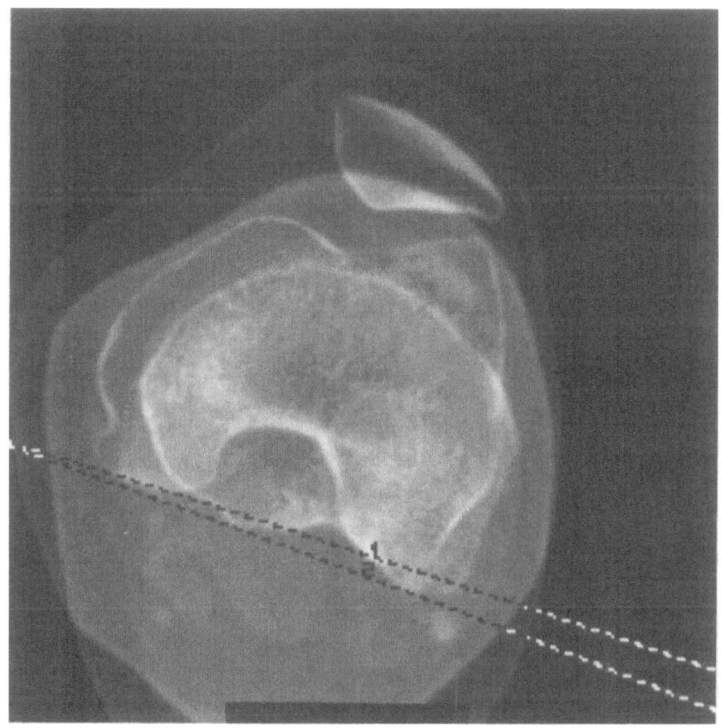

Rotation of the knee. This image is an addition of two scans (see text). A line *(1)* joining
the posterior aspects of the two condyles is drawn. A second line *(2)* tangential to the
posterior part of the tibia is also drawn. The angle between these two lines corresponds
to the rotation of the knee, in this case 5°. This angle is rarely measured, generally only
when the torsion of the whole lower extremity is to be assessed

is advisable to immobilize the thighs and the lower legs with straps. As the FPG-TT distance should be measured in a plane perpendicular to the mechanical axis, foam pads should be put between the knees and the ankles. The feet should not touch, as this would cause adduction of the leg.

Thin scans (2 mm) are taken using a high-resolution bone algorithm. It is important to select the right scans. At the tibial tuberosity, the scan should pass through the insertion of the patellar tendon at the most ventral level. In the femur, the scan should pass through the superior third of each condyle, at which level the fossa between the condyles is rounded and has the shape of a Roman arch. If the scan is too proximal, the posterior fossa loses this shape. The scans are added by the CT scanner. The distance to be measured is the distance between two lines perpendicular to the bicondylar plane, one passing through the lowest part of the groove of the facies patellaris femoris and the other through the tuberositas tibiae (Fig. 4). Normal values range from 5 to 13 mm, with a mean value of 9 mm.

Knee Rotation

Knee rotation, although not of very great interest, is easy to measure using CT. Two scans have to be taken, one at the level of the posterior aspects of the femoral condyles and the other at the level of the tibia plateau. These two scans are added and on the resulting image one line is drawn joining the posterior aspects of the two femoral condyles, another across the posterior aspect of the tibial plateau. The angle between these lines corresponds to the knee rotation (Fig. 5).

Torsion of the Tibia

Torsion of the tibia is also easily measured. One scan is made at the superior part of the tibia just above the tibiofibular joint, and another at the level where both malleoli and the talus are seen. These two scans are superimposed. Two lines are drawn on the resulting composite image, one at the level of the posterior aspect of the tibial plateau and the other joining the centers of the malleoli and the center of the talus. This second line should be parallel to the anterior and posterior borders of the talus. The angle between these two lines corresponds to the torsion of the tibia (Fig. 6).

The Length and Axis of Bones

Measurements of bone length are quite accurate when using the digital imaging facilities (topogram or scout view) of the CT, as table movement can be controlled precisely. The measurements are accurate as long as they are performed parallel to the long axis of the CT table; the structures to be measured should be positioned accordingly (Fig. 7). Limb length discrepancies are easily detected. As the CT method reduces the radiation dose, it should be preferred to conventional methods. A further advantage of CT is that the X-ray beam is always perpendicular to the structure to be measured, whereas with conventional methods the structures are imaged by divergent X-rays.

Axes are easily measured on topograms or scout views. As in conventional radiography, however, one must pay attention to correct projection of the structures in order to avoid erroneous measurements.

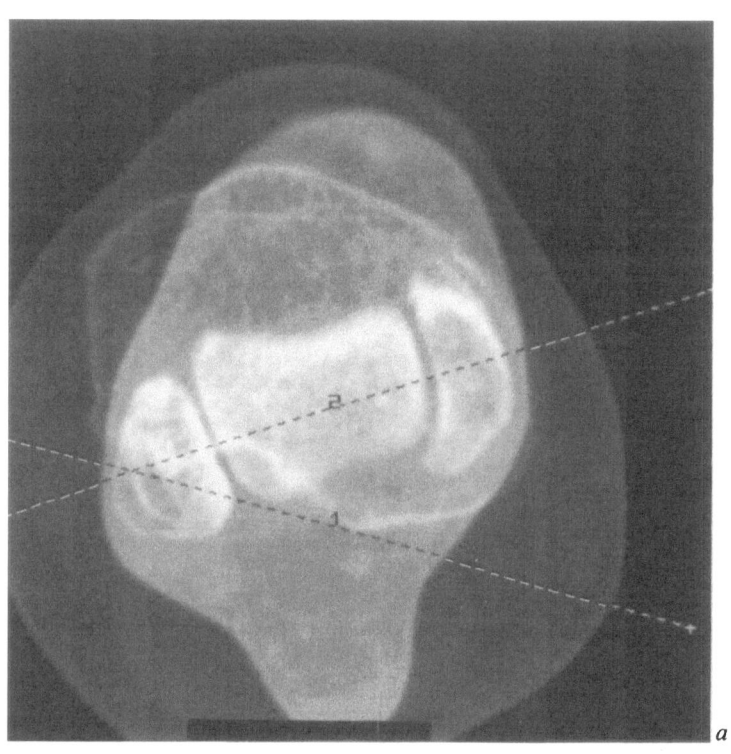

a

Figure 6 a, b

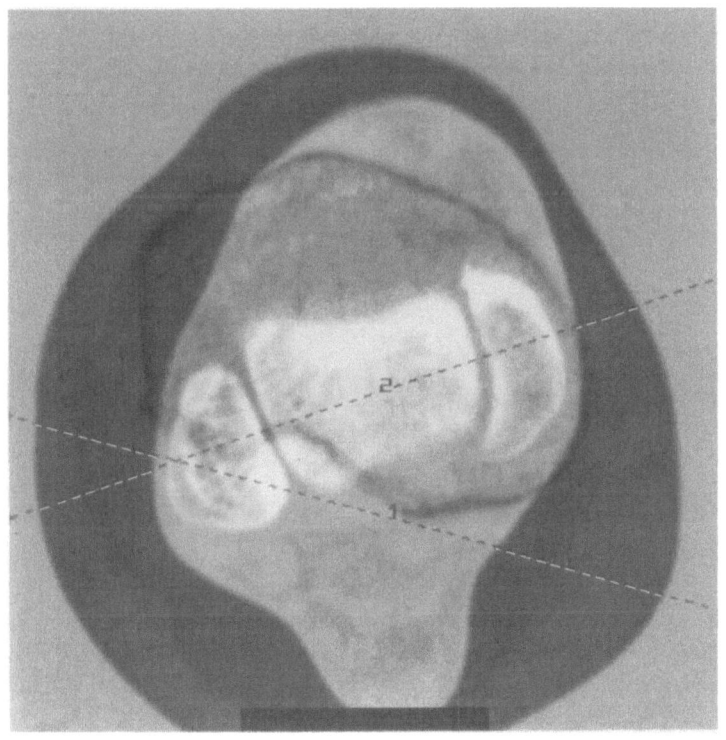

b

a Torsion of the tibia (27°). This image is an addition of 2 scans (see text). A line *(1),* is drawn tangential to the posterior part of the tibia. A second line *(2)* is drawn passing through the middle of both malleoli and as parallel as possible to the ventral and dorsal borders of the talus. The angle between these two lines corresponds to the torsion of the tibia, in this case 27°. *b* Same measurement. Instead of adding the two reference images, they have been subtracted. This technique may be useful when the structures are not well defined using the addition technique

109

Figure 7

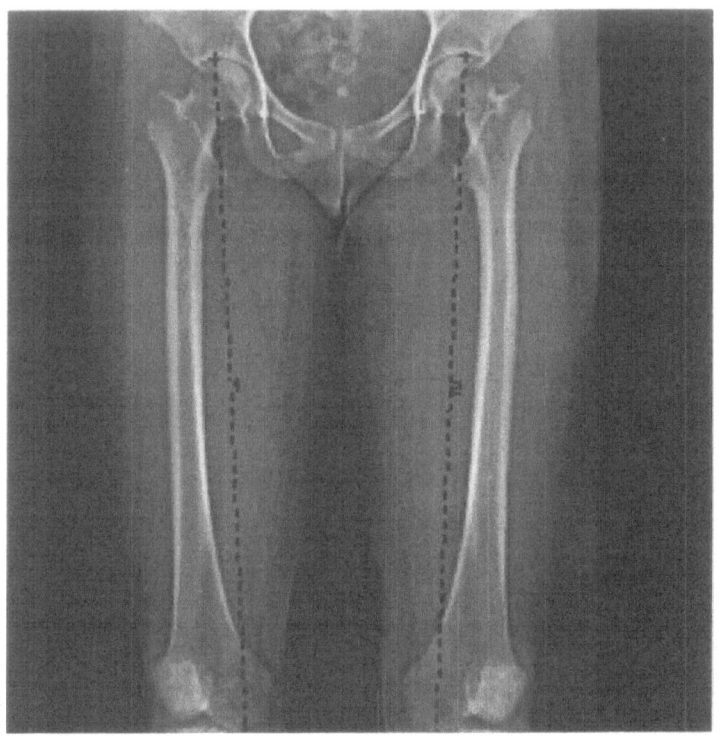

Length measurement. The structures to be measured should be parallel to the long axis of the CT table. The reference points chosen should be the same on both sides and should not be changed on follow-up studies

Coverage of the Femoral Head

Existing Methods

Assessment of the coverage of the femoral head by the acetabulum has always been a challenge to radiologists and clinicians. A substantial lack of coverage like that in acetabular dysplasias, is easy to recognize, but borderline deficiencies of coverage are far harder to detect. The difficulty in evaluating femoral head coverage arises from the fact that the acetabulum is in neither a purely frontal nor a purely sagittal plane. The borders of the acetabulum can be recognized on conventional radiographs, but it is sometimes difficult to delineate exactly the anterior and posterior margins. A simple approach to evaluating the coverage of the femoral head is to measure the CE angle (Fig. 3). An angle above 25° is normal and one below 15° is pathological. The CE angle merely gives an appreciation of the lateral coverage; it yields no information about the anterior coverage or the extent of the acetabular notch. Another approach, using CT, involves addition of scans, one at the maximal diameter of the femoral head and the other at the top of the femoral head. This method is also approximate, as it only evaluates the situation in two planes. The only accurate method is assessment of the cartilaginous surface of the acetabulum. Currently, CT scanners have no suitable software, but some such measurements using CT data have been performed on mainframe computers for research purposes. We were looking for a way of performing such measurements on less expensive and more widely distributed hardware, and wrote the necessary software ourselves.

Programs on the Disk

We (H.-M. Hoogewoud and P. Cerutti) designed and wrote, in Turbo Pascal V.5.00 (Borland), the programs necessary for evaluation of the weight-bearing surface of the hips. These programs are on the disk provided with this book.

Hardware Required

The major requirement of the program design was that the software run on IBM PCs and compatibles, as these are readily available and not too expensive. Neither a workstation nor a mainframe computer should be necessary. The programs written, then, run on IBM PCs or compatibles with 640 KB RAM and a graphic monitor resolving 640×200 (CGA), 640×350 (EGA), 640×480 (VGA) or 640×400 (AT & T or Olivetti standard) pixels. Also required are a graphic tablet (CRP or Summagraphics bit pad one compatible) for the data input and a printer with text and graphic capacities (compatible with Epson standard) for hardcopies of the three-dimensional reformatting of the acetabular cartilaginous surface.

Data Sources

As we wanted the programs to run on PCs using data from all the different existing scanner types, we used the film hardcopies as data sources for the measurement of the coverage of the femoral head. The variable floppy disk sizes and formats and the different archiving systems of the CT scanners are generally not IBM-compatible. It would have been too complicated to interface our software with all the existing CT types.

Scanning Technique

In order to reduce the partial volume effect and to increase the precision of edge delineation the scans should be 2 mm thick. As only contour data are required, mAs settings may be low. In order to obtain a high degree of precision the scans should be contiguous at the upper fourth of the femoral head (Fig. 8). More caudally, the table feed may be 4 mm until no more acetabular cartilage is seen. Each hip is recalculated separately with a high-resolution bone algorithm and enlarged so that the acetabulum and the femoral head approximately fill the image.

What To Measure?

Ideally one should measure the surface of the virtual space between the cartilage of the femoral head and the cartilage of the acetabulum. The

Figure 8

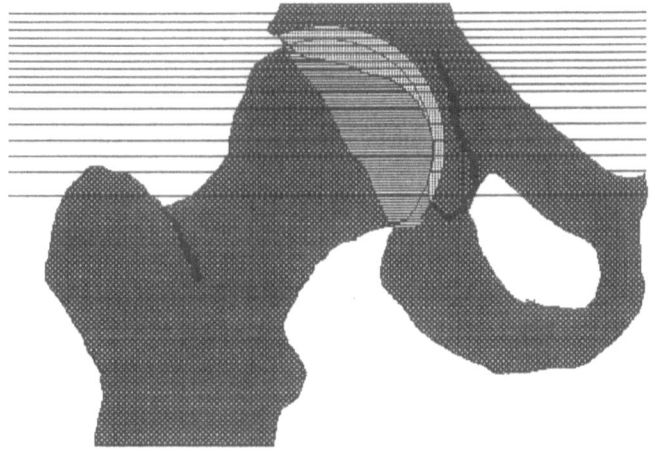

Scanning technique for morphometry of the hip. Because of the partial volume effect, scans should be thin (2 mm). High precision is needed at the top, so the table feed there is 2 mm. In the lower three fourths of the femoral head the table feed may be 4 mm. Because only contour data are needed, the mAs setting may be low

exact border between acetabular cartilage and femoral cartilage can not be seen on plain scans. Arthrography would show this delimitation, but we wanted our method not to be invasive. We thus assumed the border to be halfway between the bony edges of the acetabulum and the femoral head. Arthrographies in other patients showed this assumption to be reasonable.

Data Sampling

The contour data for the morphometry program are successively gathered from each simple scan. Each scan should occupy exactly the same position on the graphic tablet. On the first scan, the coordinates and a scale factor, which will remain the same for all the scans, have to be defined. With the electronic pen the contour of the assumed border between the acetabular and the femoral cartilage is followed successively counterclockwise on each scan. This direction of data sampling is important, as the software uses it to determine the spatial orientation of the surfaces. The morphometry program allows variable resolution, but typically 24 edge points per scan are selected. When all the data have been entered, the morphometry program will join all the edges of the different planes using a segmentation algorithm, creating finite elements (triangles). The series of triangles whose edge coordinates, position, surface and orientation in space are known is computed by the software.

Data Processing

The orientation and the magnitude of the load vectors acting on the hip joint were calculated by Pauwels (1973) and more recently by Maquet (1985), for the different phases of gait as defined by Braune and Fischer (1895).

As the orientation of each single triangle is also known, values for the following are computed in the different phases of gait (Tables 1, 2):

— *The weight-bearing surface.* This corresponds to the sum of the surfaces of the single triangles whose angle between normal vector and load vector is less than 90°. Non-weight-bearing triangles are thus eliminated.

— *The weight-bearing projected surface.* This corresponds to the sum of the projected surfaces of the weight-bearing triangles (as defined

114

Table 1. Normal surface and load data at three different phases of gait ($n = 40$ hips). Variation in load is much greater than variation in coverage of the femoral head

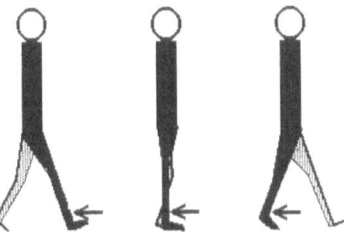

		Phase 12	Phase 16	Phase 22
Real surface (mm^2)	mean	1846	1824	1781
	SD	301	309	311
Projected surface (mm^2)	mean	1217	1199	1175
	SD	193	194	192
K (load factor)		4.32	1.83	3.40
Load (kg/cm^2)	mean	24.8	10.7	20.2
	SD	2.80	1.23	2.35

Table 2. Computer printout from the morphometry software. The coverage of the femoral head (in mm^2) is indicated at six different phases of gait. *Realsurf* is the sum of the surfaces of each single triangle whose angle between normal vector and load vector is less than 90°. *Projsurf* is the sum of the projected surfaces (on a plane perpendicular to the load vector) of the same triangles

```
HIP 3D MORPHOMETRY
==================

Raw data File Name : CT6225R.RDT    RIGHT HIP
---------------------


Surfaces at different gait phases :
-----------------------------------

    phase : 12  realsurf: 1662  projsurf: 1098
    phase : 14  realsurf: 1622  projsurf: 1098
    phase : 16  realsurf: 1601  projsurf: 1090
    phase : 18  realsurf: 1607  projsurf: 1087
    phase : 20  realsurf: 1592  projsurf: 1074
    phase : 22  realsurf: 1581  projsurf: 1071
```

Figure 9

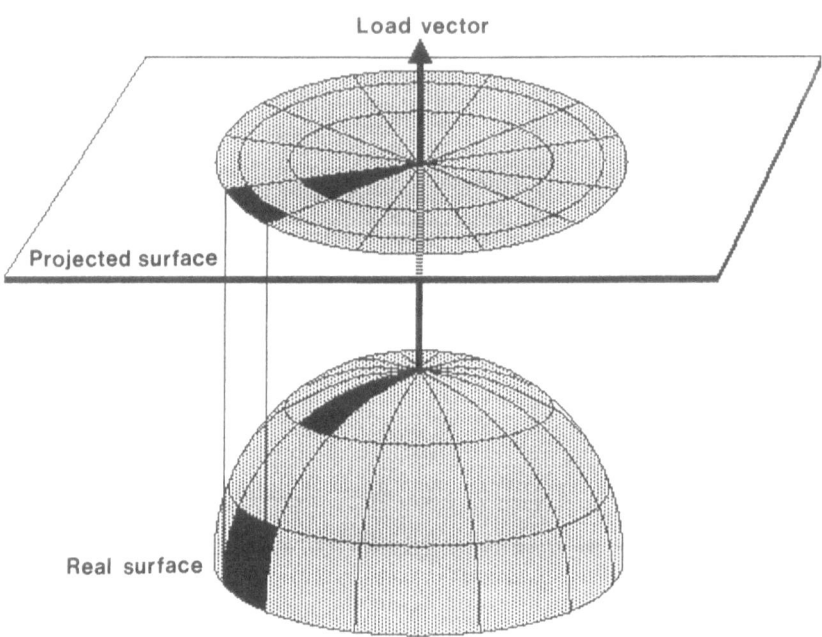

The real surface of a hemisphere and its projection on a plane normal to the load vector

above) on a plane normal to the load vector, and is generally used to determine the articular compressive stress *(CS)* in a congruent ball and socket joint (Fig. 9).

$$CS = \text{force/projected surface}$$

From the above assessed surfaces, values for the following can also be determined:

— *Percentage of coverage of the femoral head (PCF)*. This measure of how much of the upper half of the femoral head is covered by ace-tabular cartilage is derived by dividing the weight-bearing surface by half the surface of a sphere with a diameter corresponding to the maximal diameter *(D)* of the femoral head, the femoral head being considered as a sphere for the sake of simplicity.

$$PCF = \text{weight-bearing surface}/(D^{2*}\pi/2)$$

— *Articular stress*. As the weight of the patient and the magnitude of the load vectors are known, it is possible to calculate the pressure on the cartilage in kg/cm^2 *(CS)*

$$\text{force} = \text{weight} * K$$
$$CS = \text{force/projected surface}$$

K is an acceleration factor depending on the phase of gait (Table 1).

Normal Values

Normal values were obtained using our software from 40 hips of 30 asymptomatic adult patients, mean age 35 years. The results are shown in Table 1 and Fig. 10.

Figure 10

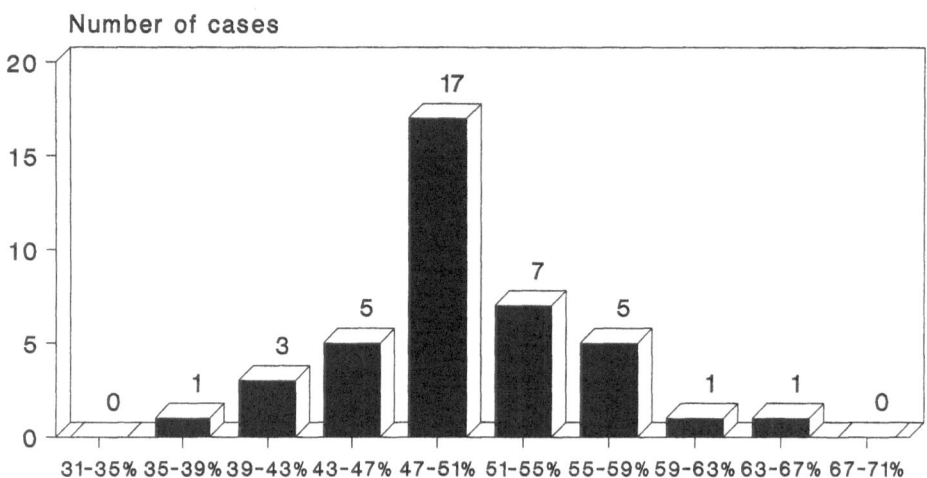

Distribution of the percentage of femoral head coverage in 40 normal adults hips. Mean coverage 50%, SD 5.33%

Three-Dimensional Graphics Software

The second program we wrote was a graphic program allowing the drawing of objects in three dimensions using hidden line and hidden surface algorithms as described by Newell et al. (1972). The program is able to draw a variety of objects (in Fig. 11, a church and a model of the femur), provided the data to be used are written in the right format. Besides performing the various measurements, the morphometry program produces data files for the graphic program. The spatial delineation of the coverage of the femoral head may thus be analyzed (Fig. 12). Zooming, rotation, shading, and differentiation of the inner and outer aspects are all possible.

Comments

- The lack of coverage of the femoral head and overweight, both resulting in an increase in articular stress, are only partial causes of osteoarthritis. The quality of the cartilage, for instance, cannot yet be assessed with precision. Future research may show the importance of each factor.
- Our series of normal hips showed poor correlation between the CE angle and the percentage of coverage. This seems to be due mainly to the importance of the medial notch. In some patients with a normal CE angle it could be shown that the low degree of coverage was due to a very large medial notch. The acetabular cartilaginous surface resembles not, as one might imagine, a hemisphere, but rather a strip wrapped around a hemisphere.
- The methodology can be used for planning corrective surgery, particularly periacetabular osteotomy. Articular stress and weight-bearing surface can be recomputed after virtual displacement of the acetabulum. Software providing this facility is currently in development.
- Further details about the software can be found on the *.doc files on the floppy disk enclosed.

119

Figure 11

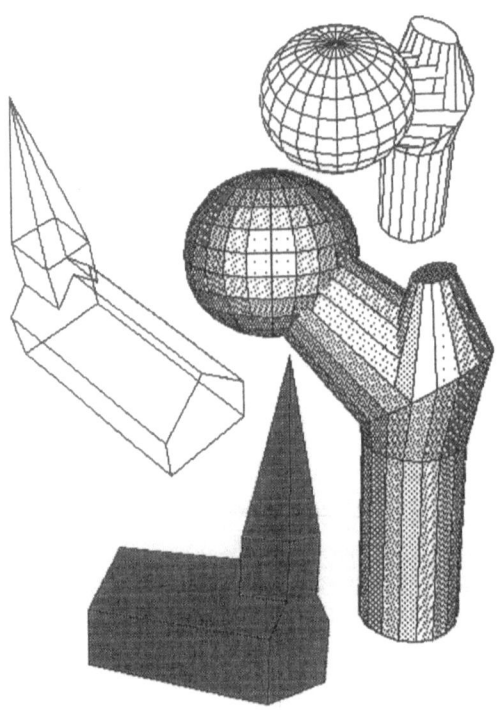

Different drawings obtained using the three-dimensional graphics software

Figure 12

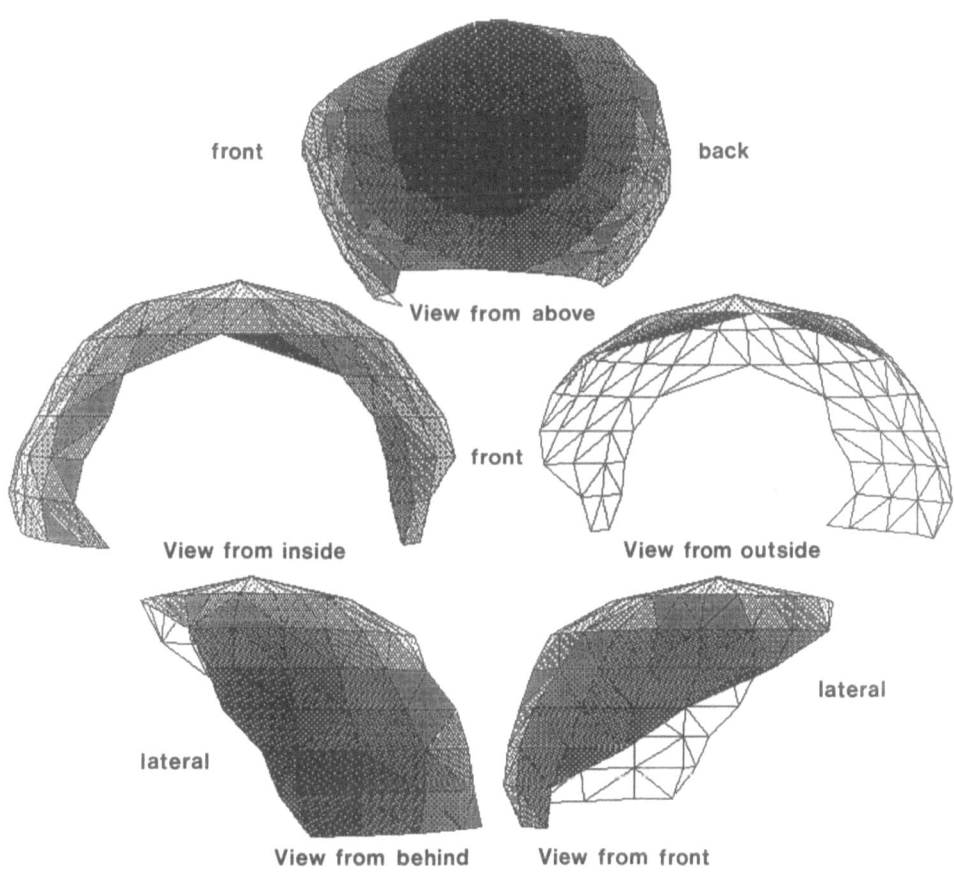

Different views of the cartilage of an acetabulum. The data were obtained by the morphometry software and used by the three-dimensional drawing software to produce
these spatial representations. Note the importance of the medial notch

References

Benninghoff (1985) Anatomie. Urban & Schwarzenberg, Munich (3 vols)

Braune W, Fischer O (1895) Der Gang des Menschen. I: Versuche am unbelasteten und am belasteten Menschen. Abhand. der math.-phys. Cl. d. k. Sächs. Ges. der Wissensch. 21: 153–322

Gambarelli J, Guérinel G, Chevrot L, Mattèi M (1977) Ganzkörper-Computer-Tomographie. Springer, Berlin Heidelberg New York

Hafferl A (1969) Lehrbuch der topographischen Anatomie. Springer, Berlin Heidelberg New York

Hafner E, Meuli HC (1975) Röntgenuntersuchung in der Orthopädie: Methode und Technik. Huber, Bern

Hounsfield GN (1973) Computerized transverse axial scanning tomography. Part I: Description of system. Br J Radiol 46: 1016–1022

Lang J, Wachsmuth W (1972) Bein und Statik. Springer, Berlin Heidelberg New York (Praktische Anatomie, vol 1/4)

Maquet PG (1985) Biomechanics of the hip. Springer, Berlin Heidelberg New York

Newell ME, Newell RG, Sancha TL (1972) A solution to the hidden-surface problem. Proceedings of ACM National Conference, August

Nomina anatomica, 5th edn. (1983) Williams and Wilkins, Baltimore

Pauwels F (1973) Atlas zur Biomechanik der gesunden und kranken Hüfte. Springer, Berlin Heidelberg New York

Peterson RR (1980) A cross-sectional approach to anatomy. Year Book, Chicago

Rauber, Kopsch (1988) Topographie der Organsysteme, Systematik der peripheren Leitungsbahnen. Thieme, Stuttgart (Anatomie des Menschen, vol 4)

123

Selected Reading

Radiology

Dunn DM (1952) Anteversion of the neck of the femur. J Bone Joint Surg 34B: 181–186

Helms CA, McCarthy S (1984) CT scanograms for measuring leg length discrepancy. Radiology 151: 802

Hernandez RJ, Tachdjian MD, Poznanski AK, Dias LS (1981) CT determination of femoral torsion. Am J Radiol 137: 97–101

Jakob RP, Haertel M, Stüssi E (1980) Tibial torsion calculated by computerized tomography and compared to other methods of measurement. J Bone Joint Surg 62B (2): 238–242

Morvan G, Massare C, Frija G (1986) Le scanner ostéoarticulaire. Techniques d'utilisation, indications, résultats. Vigot, Paris

Murphy SB, Simon SR, Kijewski PK, Wilkinson RH, Griscom NT (1987) Femoral anteversion. J Bone Joint Surg 69A: 1169–1176

Rippstein J (1955) Zur Bestimmung der Antetorsion des Schenkelhalses mittels zweier Röntgenaufnahmen. Z Orthop 86: 345–360

Biomechanics

Braune W, Fischer O (1889) Ueber den Schwerpunkt des menschlichen Körpers mit Rücksicht auf die Ausrüstung des Deutschen Infanteristen. Abhand. der math.-phys. Cl. d. k. Sächs. Ges. der Wissensch. 15: 560–672

Greenwald AS, Haynes DW (1972) Weight bearing areas in the human hip joint. J Bone Joint Surg 54: 157–153

Kummer B (1978) Anatomie fonctionnelle et biomécanique de la hanche. Acta Orthop Belg 44: 94–104

Kummer B (1979) Die Tragfläche des Hüftgelenks. Z Orthop 117: 411–712

Computer Science

Bret M (1988) Images de synthèse. Méthodes et algorithmes pour la réalisation d'images numériques. Dunod informatique. Bordas, Paris

McGregor J, Watt A (1986) The art of graphics for the IBM PC. Addison-Wesley, Wokingham, GB